NIST Technical Note 1764

High-Temperature Guarded-Hot-Plate and Pipe Measurements: 2nd Operators Workshop (March 19-20, 2012) Co-sponsored by ASTM Committee C16 on Thermal Insulation

Robert Zarr
Engineering Laboratory
Energy and Environment Division

Thomas Whitaker
Industrial Insulation Group

Frank Tyler
Owens Corning

http://dx.doi.org/10.6028/NIST.TN.1764

November 2012

U.S. Department of Commerce
Rebecca Blank, Acting Secretary

National Institute of Standards and Technology
Patrick D. Gallagher, Under Secretary of Commerce for Standards and Technology and Director

Table of Contents

1. Introduction

The workshop on high-temperature guarded-hot-plate and pipe measurements was held on March 19-20, 2012 at the National Institute of Standards and Technology (NIST) in Gaithersburg, Maryland, USA. The workshop was co-sponsored by the ASTM International Subcommittee C16:30 on Thermal Measurement and by the National Institute of Standards and Technology. This was the second workshop in a series that specifically focused on the needs of operators conducting these types of measurements.

2. Workshop Objective and Goals

The objective of the workshop was to examine and to improve the general understanding of the operation of the guarded-hot-plate and pipe apparatus at elevated temperatures (up to 650 °C).

The ultimate goal of the workshop was to examine and, hopefully, to reduce the present levels of variation in inter-laboratory comparisons of guarded-hot-plate and pipe apparatus measurements at elevated temperatures. To support this effort, five general areas of discussion were identified by the organizers and are presented below:

1) the role of NIST with respect to the development of high-temperature thermal insulation reference materials and/or measurement services for the public;
2) the role of calibration in the metrological traceability for primary input (and some secondary) quantities that are required for the determination of steady-state thermal transmission properties;
3) the need for effective and accurate control strategies for the guarded-hot-plate and pipe apparatus;
4) the assessment of measurement uncertainties by using international guidelines such as the Guide to Expression of Uncertainty in Measurement (GUM) and,
5) the development and use of design of experiment approach in the planning of future inter-laboratory comparisons.

3. Participants

The participants of the workshop included representatives from the following general categories: thermal insulation producers that utilize the equipment, equipment manufacturers, testing laboratories, consultants, academics, and government. Exhibit 1 presents a breakdown of the participants. Appendix A is a directory of the participants. There were seven international participants from Canada and Europe.

Exhibit 1. Breakdown of workshop participants by user category

Category	Participants
Thermal insulation manufacturer	9
Equipment manufacturer or testing laboratory	7
Consultant	2
Academic	1
Government	8
TOTAL	27

4. Agenda

The workshop was organized into four sequential sessions. The technical approach was based on the 1st Workshop Proceedings (Appendix B), which noted that the plate and pipe methods, although different in physical geometry, confront many of the same technical challenges. As a result, this workshop was organized along common generic issues. Each session included presentations followed by an open discussion period. The discussion minutes were recorded by the Session Secretary, edited by the Workshop Organizers, and described in Section 5. The presentations for Session 1 through Session 4 are represented with permission of their authors in Appendix C through Appendix F, respectively.

4.1. Session 1

Session 1 focused on the current status of high-temperature thermal insulation reference materials in North America and Europe. An overview of NIST research activities in this area were given as well as an overview of the NIST Standard Reference Data, Standard Reference Material, and Measurement Service Program.

Session 1 (Day 1)*			
Moderator:	**Bill Healy**	**Secretary:**	**Monyelle Mingo**
9:00	Welcome and Introductions		Bill Healy
9:15	Workshop Overview		Robert Zarr
A. High-temperature Reference Materials and Measurement Services			
09:30	Industry needs, ASTM perspective		Tom Whitaker
09:50	Europe outlook		Erik Rasmussen Roland Schreiner
10:05	BREAK – 10 min.		
10:15	Summary of NIST research		Robert Zarr
10:30	Overview of NIST SRM & Measurement Services		Robert Watters
11:00	DISCUSSION		
12 Noon	LUNCH (NIST Cafeteria)		

*Building 224, Room B245 unless otherwise stated

4.2. Session 2

Session 2 provided tours of the NIST metrology laboratories for electrical and dimensional measurements located in the Advanced Metrology Laboratory Complex (AML). (Unfortunately, the temperature metrology laboratories were unavailable due to a schedule conflict with the 9th International Temperature Symposium.) An overview of research activities at the Laboratoire national de métrologie et d'essais (LNE) were presented. The presentation was followed by a tour of the NIST guarded-hot-plate facilities which includ-

2

ed a short presentation on the construction of the 500 mm guarded-hot-plate apparatus and a preview of the PID control strategies for the apparatus. Summary discussions were organized afterwards on topics identified in the 1st workshop including thickness measurements, sensor reliability and accuracy, guard imbalance, and surface emissivity.

Session 2 (Day 1)*			
Moderators:	**Tom Whitaker, Frank Tyler**	**Secretary:**	**Leah Strohsnitter**

13:00 Tour of NIST Metrology Laboratories (AML)
Electrical (Building 218, Room F0013) Rand Elmquist, Richard Steiner, Yi-hua Tang
Dimensional (Building 219, Room G024)   John Stoup

14:30 Laboratoire national de métrologie et d'essais (LNE) GHP   Alain Koenen
(Building 224)

15:00 Tour of NIST GHP Facilities (Building 226)   Robert Zarr, Bill Healy
(advance demonstration of PID computer model)   Bill Thomas

B. Thickness Determination
16:00 DISCUSSION   Attendees
Suggested topics:
    Plate: use of spacers
    Pipe: pin versus circumference methods, bands
    Temperature effect (measure or use literature values)

C. Sensor Reliability and Accuracy
16:20 DISCUSSION   Attendees
Suggested topics:
    Uniformity check: number of sensors, locations
    Type of sensor: thermocouple versus platinum resistance thermometer
    Thermocouples: sheathed versus non sheathed (grounding issues)
    Temperature effect – change with cycling, degradation (how to check)

D. Guard Imbalance Check
16:40 DISCUSSION   Attendees
Suggested topics:
    Uniformity check: number of sensors, locations
    Thermocouples versus thermopile
    Imbalance experiment
    Gap insulation, emittance, expansion

E. Surface(s) Emissivity
17:00 DISCUSSION   Attendees
Suggested topics:
    How to measure
    Coatings – emittance, durability

17:30 Adjourn Day 1

18:00 Dinner at local restaurant (optional – reservations requested)   Return to hotel

*Building 224, Room B245 unless otherwise stated

4.3. Session 3

In Session 3, there were three presentations on PID (proportional, integral, derivative) control: 1) how to model the apparatus for simulation control strategies, 2) application to the plate method; and 3) application to the pipe method. After the Discussion period, two presentations were given on uncertainty analyses with applications on specific examples for the guarded-hot-plate method.

Session 3 (Day 2)*		
Moderators: Tom Whitaker, Frank Tyler		**Secretary: Frank Tyler**
F. <u>Control System Considerations and Steady-State Issues</u>		
09:00	Simulation of PID control model (plate)	Bill Thomas
09:30	Application of PID control model (plate)	Robert Zarr
09:55	Industry issues PID control (pipe)	Tom Whitaker
10:15	BREAK – 10 min.	

10:25	DISCUSSION
	Suggested topics:
	    PID temperature control versus locking power (temperature drifts)
	    Level of control at different temperatures (how precise?)
	    How many data points per run
	    How to define steady-state
		Attendees

G. <u>Uncertainty & Reporting</u>		
11:00	Introduction to GUM	Robert Zarr
11:20	Uncertainty Analysis using the GUM and GUM supplement	Blaza Toman

11:45	DISCUSSION
	Suggested topics:
	    How to use GUM uncertainty budget to rank sources of uncertainty?
	    How does uncertainty analysis complement DEX for inter-laboratory study?
	    Introduction in ASTM C16 test methods	

12 Noon	LUNCH (NIST Cafeteria)	

    *Building 224, Room B245 unless otherwise stated

4.4. Session 4

Session 4 presented the results of two inter-laboratory comparisons. The first presentation summarized 19 years of proficiency testing in North America by NVLAP for Test Method C 177 (guarded-hot-plate apparatus) and Test Method C 518 (heat-flow-meter apparatus). The second presentation summarized two European inter-laboratory comparisons. After the presentations, the session shifted focus to future inter-laboratory needs, specifically for the pipe test method. The first of these two presentations defined needs and goals of the proposed inter-laboratory comparison. The second presentation gave a statistician's approach to the design of experiment (DEX) for a general inter-laboratory comparison with specific examples to the pipe inter-laboratory comparison under consideration.

Session 4 (Day 2)*

Moderators: Tom Whitaker, Frank Tyler   **Secretary:** Leah Strohsnitter

H.   Inter-laboratory Round Robin Recommendations

13:00	NVLAP C177/C518 Proficiency Tests	Jeff Horlick
13:20	Europe Inter-laboratory comparison	Helge Hoyer
13:40	BREAK – 10 min.	
13:50	High-temperature thermal insulation industry needs	Tom Whitaker
14:30	Experimental design (DEX) for Pipe Round Robin	Jim Filliben
17:00	Open Discussion about Workshop	Tom Whitaker, Frank Tyler
17:15	Concluding Remarks	Tom Whitaker, Frank Tyler
17:30	Adjourn Workshop	

*Building 224, Room B245 unless otherwise stated

5. Discussion Summaries

Section 5 summarizes items of interest from the question-and-answer periods, and from the discussions that were encouraged by the moderators in each session. The minutes for each session were recorded, in outline form, by the Session Secretaries (Section 4) and also by William Healy, NIST. The notes were subsequently edited by the Workshop Organizers for review by the workshop attendees. In some cases, the names of the meeting attendees have been retained for clarity. Where appropriate, action items have been noted.

5.1. Session 1

5.1.1. Industry needs, ASTM perspective

Tom Whitaker voiced concern about initiating student interest in the field of industrial thermal insulation. [This issue was re-iterated in Session 4 with a follow-up question by Robert Zarr to Professor Thomas.]

5.1.2. European perspective

Erik Rasmussen and Roland Schreiner discussed expanded glass granulate as a round robin material for thermal conductivity to higher temperatures.

- *Question 1:* Frank Tyler asked if there were any issues with dimensional stability of the glass beads. *Answer:* They are dimensionally stable. The beads are commercially available from a company in Germany. The material is inexpensive.

- *Question 2:* What about the temperature differences (delta-T) during the tests? *Answer:* On the guarded-hot-plate (GHP) tests, the delta-T is constant. For the pipe tests, the delta-T is variable.

- *Question 3:* What about the glass, is there an opacifier? *Answer:* No opacifier. There is a special method to produce an expanded glass. Optical data about the glass are available from FIW.

- *Question 4:* Is the model given in the presentation part of any standard? *Answer:* No. The true thermal conductivity takes into account conduction, radiation, convection, [etcetera].

- *Question 5:* Is there a standard for the Nusselt sphere given in the presentation? *Answer:* No. FIW is probably the only one with the sphere test method.

- *Question 6:* Can the model be standardized? *Answer:* Could circulate the Ph.D. thesis, but tough to understand.

- *Comment 1:* Five different labs have been part of the testing with glass spheres. They hope to have all labs in Europe and some in North America. The five labs have different testing sizes. They measured at 50 mm and 100 mm thicknesses. If not capable of measuring at 100 mm, measure then at the highest thickness possible. Under 20 mm, an error arises because of glass spheres. They have not measured optical effects. The delta-T for the GHP testing was 50 K. Erik Rasmussen can get more detailed results of the testing to the [workshop] group, if desired.

5.1.3. Summary of NIST research; Overview of NIST SRM & Measurement Services

Robert Zarr presented specific overview of NIST Guarded Hot Plate (GHP) activities and Robert Watters presented a general overview of the NIST SRD (Standard Reference Data), Calibration Services, and SRM (Standard Reference Material) programs. Questions and discussion were held to the end of both presentations.

- *Question 1:* Tom Whitaker asked if Europe is working to produce an SRM [using the glass spheres]. *Response:* The current effort is a private initiative.

- *Question 2:* Erik Rasmussen asked how do you determine the validity period of a reference material? *Answer* [from Robert Watters]: Some materials we know are inherently stable. Every statement on period of validity or expiration date includes the caveat that the SRM is stored and used as described in the paragraph on storage and use.. We do, on most occasions, use previous samples in the development of new reference materials. In other words, we may use a current SRM to help develop a new SRM. We do not do work to accelerate degradation. *Comment* [from Robert Zarr]: Instrument stability is important as well. To that end, the NIST GHP labs check instrument stability by establishing a measurement traceability chain for their primary sensors. The NIST GHP labs have begun a rigorous effort to send sensors back to the NIST metrology labs for control and stability checks. The stability check for the test material is more difficult.

- *Continued discussion* [from attendee]: European reference material typically has a validity period of only 5 years. Fibrous-glass boards should be (much) longer. For example, NIST certificate for 1450d has instructions for handing, storage, and use. If one follows these directions for the SRM, the certification is valid indefinitely (as stated on the 1450d certificate). Could some NIST SRM certificates have a date regarding expiration dates? If so, could these certificates extend expiration date?

- *Continued discussion:* Frank Tyler stated that ASTM Test Method C 518 requires that the SRM (or other transfer standard used for calibration purposes) be replaced every 5 years. There has been an effort to remove this text from Test Method C 518 but negative votes on the item have been found persuasive and the text has remained. The essential point in the negative votes is a request for any laboratory to demonstrate, with actual data, that the 5-year limit is either valid or invalid. Currently, no one has such data. *Action Item:* This discussion topic was identified as an action item (for the ASTM C 518 task group to collect such data).

- *Continued discussion:* Andrzej Brzezinski asked how to resolve the apparent contradiction given in the SRM [1450d] certificate and the 5-year text given in Test Method C 518. *Response:* It was stated that the Test Method C 518 document was probably written assuming that the reference materials were not always handled in accordance within their Instructions for use. *Comment:* Tim Rasinski noted that, as part of the NVLAP (National Voluntary Laboratory Accreditation Program) accreditation process, the laboratory does not need to discard [the reference material] at end of an expiration date. The NVLAP Assessor should not write a nonconformity against expired reference material as long as the lab has some process to show the reference materials have not drifted out of tolerance. The requirements of specific test methods will supersede this, however. One method to check [stability] could be to have two different units checked in two [different] devices.
- *Question 3:* Andrzej Brzezinski asked if there was a possible timeline for services by NIST. *Answer:* Robert Zarr responded that, depending on the outcome of the proposed inter-laboratory comparisons, the next stage of service from 20 °C to 200 °C is scheduled for the 2012 calendar year. The next temperature level would be scheduled afterwards.

5.2. Session 2

5.2.1. Tour of NIST Metrology Laboratories (Advanced Measurement Laboratory Complex)

Rand Elmquest, Richard Steiner, and Yi-hua Tang provided tours of the electrical resistance and voltage measurement laboratories located in the Advanced Measurement Laboratory Complex. Afterwards, John Stoup provided a tour of a dimensional measurement laboratory for the evaluation of coordinate measuring machines.

5.2.2. Laboratoire national de métrologie et d'essais (LNE) GHP – Situation in France

Alain Koenen presented an overview of current GHP activities at LNE.

- *Question 1:* For the inter-laboratory tests conducted at room temperature, was the variation of thickness known? *Answer:* No, the level of variation in the thermal conductivity data was not expected beforehand.
- *Comment 1:* In conjunction with the low-temperature thermal conductivity data presented, it was mentioned that NIST SRM 1450b was characterized to lower temperatures, down to liquid nitrogen temperatures (100 K).

5.2.3. Tour of NIST GHP Laboratory Facilities

Robert Zarr and John "Rusty" Hettenhouser presented a short photographic narration of the construction of the NIST 500 mm guarded-hot-plate apparatus. Afterwards, the attendees separated into three groups for tours of the NIST 1016 mm guarded-hot-plate facility, the NIST 500 mm guarded-hot-plate facility, and also a preview of the PID control model to be presented later in the workshop by Professor Thomas.

5.2.4. Thickness Determination Discussion (Tom Whitaker and Frank Tyler, Moderators)

As part of this topic and subsequent discussions on the primary measurements, the moderators began the dialogue with an open request for information on how attendees used spacers for their guarded-hot-plate apparatus. Subsequent questions addressed thermal expansion issues.

- *Question 1*: How do you keep your plates spaced? *Response:* Marinite[1] structural insulation machined to 0.001 in.[2] (0.03 mm) using drop gauge.

---

[1] Certain commercial entities, equipment, or materials may be identified in this document in order to describe an experimental procedure or concept adequately. Such identification is not intended to imply recommendation or

- *Question 1a*: Do you change thickness as temperature increases or do you assume constant? *Response*: Assumed to be constant.

- *Question 1b*: Does test material have different thermal expansion, if so what happens? *Response*: Usually, the test material is compressible and preconditioned. If you try to correct for [test] thickness, you would have to know what the thermal expansion [of the test material] is. If you don't know, base it [test thickness] on room temp thickness.

- *Question 1c*: Are you going to change thickness calculation based on movement? *Response:* Average data over portion that is stable.

- *Question 2*: Do other attendees do anything different? In other words, are there any other thoughts or methods to be discussed?

  - *Response*: [In Section 7, ASTM C 177 states that, for rigid and high conductance specimens, the specimen surfaces should be made flat and parallel.] In actuality, the specimen surfaces will never be 100 % parallel. European standards warn against using the average [of thickness measurements] because you need full contact. It is important to use the separation between plates. The thermal resistance measurement is between flat surfaces so if they are off your measurement is off as well. You will have gaps, need to know that you will have good contact. Biggest source of mistakes.

  - *Comment* from Tom Whitaker: For calcium silicate (rigid material), users can't measure repeatable measurements, [inter-lab results show that we] got different thickness measurements.

- *Question 3*: Are there other ways that the attendees use to control the thickness of their plates, other than spacers, and that they consider to be more reliable? *Response:* For thermal conductivity measurements at 25 °C, the plates are stopped at whatever height you want. *Comment #1:* But if [the test specimen is] compressible, you may modify density. *Comment #2:* Characterize [the test] material at target density and then decide whether or not to use spacers.

- *Question 4*: Does everyone neglect to make corrections for [thermal] expansion [for the spacers]? Is it the norm to use room temperature thickness? *Response:* Most [attendees] agree that they use room temperature [thickness]. [The] expansion [effect is limited] up to 1% for temperatures up to 650 °C.

- *Question 5*: What about for high temperature? Do we make corrections for thermal expansion effects? *Response:* This is a source of error. *Comment #1:* If you make correction for thickness you have to change density. *Comment #2:* The NIST 500 mm GHP thickness system has capability to measure [in-situ thickness] during the test, but most people do not have this capability.

- *Question 6*: What about spacers that will experience thermal expansion and therefore change [the plate] spacing? *Response:* The specimen is expanding, not the spacer. *Another response:* What if both expand?

- *Question 7:* For writing [documentary] standards, what should we do? *Response:* Have a system that you can preload and that you can measure what's actually going on like NIST's system. Then you can calibrate your system (without specimen or one that you know the expansion of). Or do

endorsement by the National Institute of Standards and Technology, nor is it intended to imply that the entities, materials, or equipment are necessarily the best available for the purpose.

[2] It is the policy of the National Institute of Standards and Technology to use the International System of Units (SI). However, in the North American construction and building materials industries, certain non-SI units are used and are reported in this publication (with SI equivalent units) to avoid confusion.

calibration with spacing material. Then you must assume the specimen will not expand more than spacer.

5.2.5. Sensor Reliability and Accuracy Discussion (Tom Whitaker and Frank Tyler, Moderators)

[*Foreword to this Discussion:* Dr. Dean Ripple, formerly with the NIST Thermometry Group, provided answers to the temperature questions from the 1st Workshop in Granville, Ohio.]

- *Discussion* on RTDs (resistive temperature devices) versus type K thermocouples (fine wire in metal sheath): One attendee noted that type K thermocouples "drifted." Another attendee suggests [using] type N thermocouples because they don't drift as much. Robert Zarr states that NIST uses type N thermocouples in thermally insulated metal sheaths. As a result of the fabrication process during vacuum brazing, the type N thermocouples were exposed to temperatures (near 925 °C) well above operation temperatures.

- *Question 1 for NIST*: Over the long term can you check the ones [type N thermocouples] that are installed? No, they are brazed in [the plates and edge-guard rings]. [However,] if you know how stable the PRT is, you should be able to measure drift [of the adjoining thermocouples in the plate].

- *Question 2 for attendee*: What kind of thermocouples do you use? Type K on previous plate systems. In newer plate systems, 100 ohm PRTs – moved away from thermocouples all together. *Question 2a*: How many [thermocouples] in the plates in the main heater (say 12 inch plate)? *Response:* For one plate system there are 3 in the main and 2 in the guard. For another plate system, there are 9 in the main heater and 8 in the guard. Again, these plate systems use PRTs, not thermocouples.

- *Same Question (2) for other attendees*: What kind of thermocouples do you use?
  - *Response 2b*: Type N, shunted thermocouples. Type K is not stable for long, so they switched to type N. Use 5 thermocouples [averaged] for one plate's temperature.
  - *Response 2c*: Type E for temperatures up to 500 °C; 4 thermocouples on each side of plate (2-sided plate). *Question from a participant:* Why do you use that type? *Response:* Highest output at lowest temperatures. But above 500 °C, you're 'pushing it'.
  - *Response 2d*: Type K, periodically push to 700 °C. At this temperature, a type K thermocouple drifts quickly.

- The moderators asked the attendees if there were other comments or questions to be discussed.
  - *Question 3 for attendees*: How long/often do you recalibrate? *Response 3a:* Some thermocouples cannot be removed, so whole probes are replaced. [In general, one attendee noted that the temperature measurement systems are re-built after a certain time period determined by the user.] *Response 3b:* Some participants have applied external calibrated thermocouples which are used as reference for internal thermocouples calibration. The external thermocouples are applied at the same position as the internal mounted thermocouple. *Question from attendee:* How can you be sure your plate is uniform? How can you get external thermocouple on the outside? *Response:* The external thermocouple is placed in contact with the plate surface (on the outside) at the same location as the plate thermocouple.
  - *Question 4 from an attendee*: Are there any issues with RTD's breaking by excessive thermal expansion? *Response:* No, they are strong. The sensors are in a metal sheath.

- It was suggested that [ASTM] could collect data on what [temperature] sensors were used, how thickness measurements were taken, how often and by what means do you calibrate, etc. and determine how this affects measurement readings.

5.2.6. Guard Imbalance Check Discussion (Tom Whitaker and Frank Tyler, Moderators)

– *Question for attendees*: Do you control primarily on thermopile or surface thermocouples on plates?

  – *Response 1*: There are several stages; first, control temperature. After [the system is] in steady state, switch to gap thermocouples. Main [heater] is controlled by thermocouples and the guard is controlled by the thermopiles.

  – *Response 2*: Surface temperatures can give error, gap temperature and edge temperature will be different and gap temperature will be somewhere in between. *Question from attendee:* Is it significant? *Response*: We control on thermocouples, not thermopiles. It was significant enough to warrant concern for change. At 1 degree [unit unspecified], they were less than half a percent. *Question from attendee*: Did you run a guard imbalance test? *Response*: Yes, but does not have data on hand. Significance may have to do with size and thickness.

  – *Response 3*: Other imbalance test: check slope to see difference in imbalance between apparatuses. They have added correction in slope to adjust balance of guard and measuring area. *Question from attendee*: Do you offset it more? *Response*: Normally it is set to zero, but change is usually seen between plus and minus. You see imbalance between guard and area, test gives option to make adjustment. *Question from attendee*: Is test run at different temperature? *Response*: Yes, at different thickness and temperature. *Question from attendee*: Have you induced a degree change between guard and analyzed the result? *Response*: 1 degree [unit unspecified] causes about 10%. Half to one degree [unit unspecified] imbalance causes significant result. Should look into effects of ratio of metered area to plate area.

  – *Comment*: NIST conducts an imbalance check on each "new thickness specimen" received from a customer. We purposely do this as part of the uncertainty evaluation. [The gap thermopile voltage] never truly runs at zero volts, there is always some offset, usually small (on the order of a few tenths of a microvolt). Thus, there is always a small heat flow through guard gap. NIST typically combines the uncertainty in this small offset as part of the uncertainty analysis. We have discovered that, even though the gap voltage is small, the uncertainty can be large, especially as specimens get thicker (i.e., heat flow through the specimen is small relative to the heat flow across the guard gap).

5.2.7. Surface(s) Emissivity (Tom Whitaker and Frank Tyler, Moderators)

– *Question for attendees*: How important is it to get 0.8 (or above) for the plate emissivity?

  – *Response 1*: No, because normally the bulk density of high temperature products is high so that radiation through specimen is minimal. For specimens having a low density, it is important to have high emissivity of plate.

  – *Response 2*: High emittance coatings for high temperature testing have been discussed in Europe. One laboratory has proposed raising the 0.8 limit to either 0.9 or 0.95. There was, however, no clear evidence of benefit, and the value was impossible to maintain in the long run. The conclusion was that there is no need to go higher than 0.8. *Question from an attendee*: Should this issue be researched more? *Response*: Yes, Helge Hoyer noted that he has a research report on the subject. It was suggested that the attendees can contact Helge Hoyer directly for the report.

- *Response 3*: The NIST GHP laboratory uses a Gier Dunkle[1] DB-100 (ASTM E 408) to measure the near normal emittance of test samples and to check the plate emittances at room temperature. [Note: The instrument was on display during the GHP tour.] NIST uses a ceramic coating for their plate surfaces that measured 0.8 (ASTM E 408). The durability of the plate coating was the main issue. *Question for NIST*: How flat is surface after application? *Answer*: Very flat, as noted in the presentation, the vendor polishes the coating as part the fabrication process.

5.3. Session 3

5.3.1. Control System Considerations and Steady-State Issues

Bill Thomas, Robert Zarr, and Tom Whitaker presented back-to-back talks on PID control for plate and pipe apparatus. Questions and discussion were held to the end of the three presentations.

Professor Thomas described the PID objectives for the simulation model to allow faster tuning than the experiment as well as tighter control. For this discussion, the control elements and PID are *all* for center plate, not for cold sides. The model neglected gradients *within* plate as far as control surfaces. There were 9 knobs to play with; PID for each of 3 sections (main heater plus double guards). The outer guard is set at an offset of 5 μV. The conduction across the gap was modeled as follows:

$$C = \frac{mC_p}{\Delta T}$$

where m = mass and ΔT = calculated time step.

The heating rate, q was modeled as follows

$$q = \frac{V^2}{R_{el}}$$

where R_{el} = electrical resistance (Ω) of a heater element.

Knowing the power supply output, an incremental control algorithm for the simulation as well as for the apparatus. A key objective was to avoid overshoot because of the lengthy cool-down time required.

Presentation 2: Robert Zarr described the control strategies for the NIST 1016 mm GHP and for the 500 mm GHP apparatus. For the 500 mm GHP, the PRT in the center of the meter plate is used as the control sensor for the meter-plate heater. Two formulae for the discrete controller have been tried: the positional form and the incremental form. Initial attempts focused on the positional form using proportional and integral (PI) control. The control, however, was unsuccessful due to "integral wind-up," a condition in which the integral term over-accumulates error summation, and then under-accumulates, resulting in long-term oscillation. The result is theoretically removable, but NIST, after several months, was unsuccessful in removal of the oscillation. At the suggestion of co-worker, the researchers switched to the incremental form which has successfully been used in controlling the 1016 mm GHP for 20 years. For the 500 mm GHP, the scan rate has been set 60 s. The resistance ratio for each SPRT is measured using a DC bridge and converted by software to a temperature. The system takes 24 h to reach steady-state conditions. Data are collected for the last 4 h.

Presentation 3: Tom Whitaker presented control strategies developed by Industrial Insulation Group for the pipe method. Industrial Insulation Group has tried controlling two ways: thermopile versus surface-mounted thermocouples. Once the power is locked, there is some temperature drift but this is of no concern (see Discussion). Whitaker noted that the steady-state definitions per ASTM C177 and C335 need improvement. ISO 8302, he noted, is similar in that regard. Industrial Insulation Group (IIG) is exploring using alternating current (AC) for main heater. New kilowatt-hour technology (shown during NIST Electrical lab tour) may enable this and may be better at high temperatures.

Discussion

- Discussion by the attendees focused initially about permanent changes to the material with temperature, e.g., hydration/dehydration or decomposition resulting in mechanical movement such as shrinkage. Options included: 1) going up "the thermal conductivity/temperature curve," then coming down "the curve;" 2) pre-heating specimen (either in the apparatus or in another conditioning chamber), etc. Arguments, pro & con, were advanced by the attendees.

  – *One viewpoint*: Only going up the curve is valid as the material will not "see" temperatures beyond its application temperature.

  – *Counter viewpoint*: Processes are designed for temperature excursions at start-up or other periods. Therefore, the material will see higher than application temperature.

  – *Comment 1*: Ideally, measurements are conducted both up and down the thermal conductivity/temperature curve. It was noted that some materials exhibit two curves, depending on the direction; in essence, signifying that there is no unique thermal conductivity for the particular material.

  – *Comment 2*: Only going down the curve may be more representative of field application but the process is time intensive due to no active cooling to remove heat from the apparatus.

  – *Comment 3*: In some cases, this discussion is a moot point because the test point order is conducted as specified by the customer.

  – *Comment 4/Open Question*: If proceeding up the thermal conductivity/temperature curve, how long does one wait at each temperature for the binder, if present, to burn out?

  – *Comment 5*: Robert Zarr requested information on the quality of ASHRAE handbook data for industrial insulation materials and whether the data were obtained going up or going down with temperature.

  – Discussion followed on control technique: In order to minimize shocking the heaters with a power surge, IIG will ramp the set point, by software, to next value. IIG uses proportional and derivative control (PD) rather than proportional and integral control (PI). Allows [temperature] drift with [power] lock because precision is not a requirement based on overall curve. IIG uses Fuji controllers which will do auto-tuning at high temperatures [control] not so good at higher temperatures. The controller has ability to respond to software commands. Software can query controller as to the voltage output of the power supply and can switch it into constant power.

Control Issues

- New techniques: IIG is now looking at AC for main heater control.
- IIG stated that, for control, the controller requires different PID control constants for different temperature regimes.
- Several attendees discussed the requirements for the digital-to-analog (D/A) conversion. Attendees used different D/A converters having the following values: 24 bit, 12 bit, or 10 bit (among others). Some attendees noted that 10 bit and 12 bit D/A converters are unsatisfactory for precise control due to insufficient resolution capability. It was noted that 16 bit D/A are adequate, commercially available, and are currently inexpensive.
- One attendee noted that, because power is square over [electrical] resistance, if you change the control voltage by a factor of 2, the power will change by a factor of 4, which can compound control with a "hunting" problem.

- One of the issues is when you get to lower outputs because of high thickness insulation, it's hard to control very low power levels.
- Some labs use switchable in-line precision resistors for better control in low/high ranges.

5.3.2. Uncertainty & Reporting

Robert Zarr presented an introductory talk on the GUM and Blaza Toman presented an uncertainty analysis based on statistical approach recently approved in GUM supplement.

- *Question 1*: Bill Healy asked the attendees, with regard to the temperature range, what level of uncertainty are the attendees comfortable with at high temperature levels? *Response*: approximately 5% with some debate.
- *Comment 1 and Question 2*: Tom Whitaker stated he was somewhat uncomfortable about European EC Certification. He asked how do you go to a lab, get data, and know that you will meet that level? *Response*: Erik Rasmussen answered that the rules are made for consumer protection; not supposed to declare mean, but to declare curve with some kind of uncertainty. In Europe, [we are] aiming for 'safe values', not just mean, which reduces liability for the user. Uncertainty in testing is very important to Europeans.
- *Comment 2*: Tim Rasinski stated that as a NVLAP representative he is asked by participating labs, "why do we have to do an uncertainty calculation?" He stated that Germany is putting the uncertainty in the law; Europeans must use GUM and report coverage factor.
- *Comment 3*: Robert Zarr stated, should we not ask that, during an ASTM inter-laboratory study (ILS), each lab report its measurement uncertainty? He noted that this practice is currently required for inter-laboratory key and pilot comparisons among the national metrology laboratories.
- *Comment 4*: An attendee stated that one cannot attempt the uncertainty analysis to the level of commitment from NIST. Rasinski (NVLAP) stated that a laboratory need only have to consider primary contributors for the GUM, not necessarily all the sources cited by NIST (Zarr's presentation).
- *Comment 5*: The attendees stated most labs in the U.S. report a precision and bias (PB), based upon ASTM method's P&B section. Rasinski (NVLAP) stated that according to ISO 17025, if the equipment is built per prescribed ASTM method, then they can use the PB statement, but the reality is, some equipment built per spec is known to be outside the PB so there's a disconnect.
- *Comment 6*: Erik Rasmussen stated that this is an excuse (i.e., to use PB). That's why we're having this workshop. In Europe, they're demanding this type of exercise. They may skirt obligation in Europe as long as they can prove that they're 'working on the subject'.
- *Comment 7*: Rasinski (NVLAP) stated that if a manufacturer is considering shipping products outside of the U.S., you're going to have people (i.e., auditors) carefully examining your uncertainties. Rasinski cited a recent case with an auditor in Australia that examined, questioned, and ultimately rejected an uncertainty calculation from a domestic manufacturer.
- *Comment 8*: Helge Hoyer stated that we need statistics. You will always find a single item that will be below a standard, so statistics are needed not only for producers but for consumers.

5.4. Session 4

5.4.1. NVLAP C177/C518 Proficiency Tests

Jeffrey Horlick summarized 19 years of NVLAP proficiency test results for ASTM Test Methods C 177 and C 518.

- *Comment 1:* Tim Rasinski announced that NVLAP is currently seeking outside services to help run the NVLAP proficiency testing program. This individual person would need to coordinate the attainment of material, preparation of the material, and shipping of material to the laboratory participants. Mr. Rasinski also stated that NVLAP is also seeking one more individual to conduct accreditation assessments of thermal insulation products.

- *Comment 2:* It is important to note that all of the results presented in the talk by Jeffrey Horlick were at room temperature.
- ***Action Item 1:*** Robert Zarr recommended that the results of this presentation (which have been previously published in the open literature) be forwarded to ASTM Task Group C 177 and Task Group C 518 for consideration as new item of business for inclusion as part of the precision and bias statement of the respective test methods.
- *Comment 3:* Tom Whitaker noted that over the course of 19 years, we have not made much improvements in the results.

5.4.2. Europe Inter-laboratory comparison

Helge Hoyer presented results from two high-temperature inter-laboratory comparisons conducted in Europe by the European Mineral Wool Association.

- *Question 1:* Was there an attempt to plot thermal conductivity versus bulk density? *Answer:* Yes, but the results were not neat.
- *Question 2:* How much variation for the second material, 38mm? *Answer:* Four labs used 38 mm spacers; two labs 37.8 mm; and one lab 37.99 mm.
- *Question 3:* What was the range of variation? *Answer:* Some labs have fixed differences and some just use the spacers.
- *Question 4:* Given small number of labs, were there any statistical tests done? *Answer:* No.
- *Comment 1:* Looking at Lab G (on the slides), it's running low from 100 °C to 500 °C. That is a 1-in-6 chance of a lab running low. Raising that to the power of 4 will result in variation under 5 %.
- *Question 5:* The maximum deviations are all positive, does that mean anything? *Answer:* No.
- *Comment 2:* Tests are randomized; the labels A-G do not represent order.
- *Question 6:* On mineral wool round, was the material preconditioned? *Answer:* No.

5.4.3. High-temperature thermal insulation industry needs

- *Question 1:* Where can we improve round robin testing?
  - *Response 1:* Now requiring use of calipers for measurement, give actual measurement. Requiring densities to be presented along with dimension measurements and mass. Asking for raw temperature (i.e., thermocouple) data as well as average temperature of the hot and cold surfaces.
  - *Response 2:* There is a broad variation in design of equipment and lengths (diameters are the same; all are 3 in. pipes). This may have an impact.
  - *Response 3:* The test material has not been selected: difficulties with calcium silicate and with mineral wool. Is there a material that works well in a high-temp round robin to be used on a pipe apparatus that can be passed from lab to lab, etcetera? With glass piece, you must have stronger mechanical support because of weight. *Q:* How do you get the glass beads around the ends? Outside 'jacket' is put on (usually plastic). At the last ASTM meeting, a decision was made for mineral wool to get the round robin going, objections?

5.4.4. Experimental design (DEX) for Pipe Round Robin

- *Comment 1:* Comment from Jim Filliben based on the previous discussion:

- *Observation 1*: There is a large k, that is, a large number of factors that can affect the response. (In the previous Discussion [Section 5.3.2], Filliben counted the factors that were discussed and stopped counting when the tally reached 18 [← high enough].)
- *Observation 2*: n is small, that is the number of tests; and,
- *Observation 3*: sigma is unknown. Get question nailed down, specification of the problem as well.

- *Comment 2*: We probably need to do a sensitivity analysis before the inter-laboratory comparison. A what if? analysis to see what is important. Select a few primary and secondary factors. We do not fully understand how sensitive the results are to some of those factors. Prior to jumping into inter-laboratory comparison, it might be best to vet some of the issues in a sensitivity analysis.
  - *Response from attendees:* Agreed with that approach.
  - *Response from Jim Filliben:* Most common experiments here [at NIST] are sensitivity studies. Filliben typically uses 2-level fractional factorial designs; with 18 factors for this study, (2^{18}) is a lot of runs. Wants 180 numbers (not sure where he got this number), would run a $2^{(18-10)} = 256$ numbers. Run this fractional factorial experiment, 2 levels. Constructing is relatively easy. Good information about main effects, but not necessarily interactions.

- *Comment 3*: As (mean) temperature increases, there is a curve in temperature [due to radiation effects]. Can run four equally spaced points either along the *X-axis* or *Y-axis*. For large multifactor experiments, look into interactions.

- *Comment 4*: Previous ASTM round robins (in the United States and North America) have not approached this design. Replication is important for these tests; if we can afford it, replicate everywhere.

- **Action Item 2:** Importance of sensitivity analysis, always lends valuable data/information about system/experiment. Look into doing 'in-house' sensitivity analysis. (Action item for next ASTM meeting).

5.4.5. Open Discussion about Workshop (Tom Whitaker and Frank Tyler, Moderators)

The following comments and questions were raised in the open discussion.

- **Action Item 3:** Request for electronic copy of presentations from this workshop to be made available.

- **Action Item 4:** The last workshop was held in 2007 which was 5 years ago. ASTM C16:30 should consider holding more frequent workshops. *Question:* Are similar workshops [on high temperature plate and pipe methods] held in Europe? *Response from Erik Rasmussen:* One already that has focused on 10 °C, another coming up in September in Belgium. This workshop has been inspiration for further work with high-temperature systems.

- *Comment 1:* Round robin presentation is similar to the results published by Mark Albers in ASTM STP 1426. Tom Whitaker is curious how results in Alber's paper compare to results presented in this workshop. He will e-mail an electronic copy of the paper from ASTM STP 1426.

- **Action Item 5:** 5-year time interval prescribed in C518, can it be taken out?

- **Action Item 6:** Temperature sensor variations: This workshop highlighted that, for thermocouples, several different kinds are used among the workshop attendees. Is there a better way? Look into platinum resistance.

- **Action Item 7:** Steady state determination: How to determine what algorithm or criteria is best for determining when steady state is reached. How do you define that you're in control?

- **_Action Item 8:_** Is it possible to have a guarded hot plate website for posting information and discussion that may occur between other workshop sessions?

- _Question for Professor Thomas:_ As a follow-up to the Session 1 concern, Robert Zarr asked to Professor Thomas to comment on how to encourage more students to become involved in this area for study and employment. Professor Thomas noted that, quite simply, involve the faculty and the students will follow. Research grants are important. Thomas further noted that people will only pay attention when something is not working properly. No one pays attention when things are running smoothly.

- _Comment 2:_ Good pace and good content during workshop, wide range of topics.

6. Summary and Recommendations

In general, this second workshop on high-temperature guarded-hot-plate and pipe measurements was well attended with participants from North America and Europe representing industry, consulting, academia, and government. The main objective of the workshop was to examine and to improve the general understanding of the operation of the guarded-hot-plate and pipe apparatus at elevated temperatures (up to 650°C). To accomplish this objective, the intention of the workshop was to present, and to discuss, the reasons and causes of differences between laboratory measurements with the ultimate goal to reduce the differences in laboratory comparisons.

To attain these objectives and goals, the workshop was organized into four sessions. Each session covered a particular subject considered important in achieving the overall goal of reducing the differences in laboratory comparisons. Session 1 covered the government's role in providing reference materials to the public and Session 2 covered issues of metrological traceability in relation to the calibration of primary measurements for the guarded-hot-plate and pipe apparatus. Session 3 covered instrument control issues and measurement uncertainty. Session 4 presented data from two recent inter-laboratory comparisons and how to design future inter-laboratory comparisons.

The general conclusion from the attendees was that the workshop was valuable and was also motivating with regards to further work with high-temperature systems. It was further noted that the first workshop was held in 2007 (five years ago) and that ASTM C 16:30 should consider holding workshops of this type more frequently. It was also requested that the workshop proceedings be made available for electronic publication.

One of the tasks of the Session Moderators was the identification of "action items" for further attention. The remaining text in Section 6 summarizes the eight "action items" that were identified by the moderators in the group discussions (Section 5). Recommendations from the workshop organizers are included with each action item. The majority of these action items are to be forwarded to the appropriate ASTM task group or subcommittee for further discussion.

1) The results of the NVLAP presentation (which have been previously published in the open literature) should be forwarded to ASTM Task Group C 177 and Task Group C 518 for consideration as new item of business for inclusion as part of the development of a modified precision and bias statement for the respective test methods.

2) The importance of sensitivity analysis always lends valuable data/information about system/experiment. The attendees should look into doing 'in-house' sensitivity analysis. (Action item for next ASTM meeting).

3) The last workshop was held in 2007 which was 5 years ago. ASTM C16:30 should consider holding more frequent workshops.

4) Request for electronic copy of presentations from this workshop to be made available.

5) The 5-year time interval prescribed in C518: can it be taken out?

6) Temperature sensor variations: This workshop highlighted that, for thermocouples, several different kinds are used among the workshop attendees. Is there a better way? Look into platinum resistance.

7) Steady state determination: How to determine what algorithm or criteria is best for determining when steady state is reached. How do you define that you're in control?

8) Is it possible to have a guarded hot plate website for posting information and discussion that may occur between other workshop sessions?

7. Acknowledgments

The success of the workshop was a direct result of the quality of the speakers and the participants. The organizers recognize the contributions from the speakers and the insightful discussions from the participants.

Appendix A: Participants

Name	Affiliation
Franz Bauer	Exova
David Bell	Southern Research Institute
Andrzej Brzezinski	LaserComp
Rob Campbell	Netzsch Instruments, North America, LLC
Rick Dolin	Industrial Insulation Group, LLC
Jim Filliben	NIST Statistical Engineering Division
William Healy	NIST Energy and Environment Division
Jeff Horlick	NIST Standards Coordination Office
Helge Hoyer	Rockwool International
Timothy Jonas	Knauf Insulation
Alain Koenen	Laboratoire national de métrologie et d'essais (LNE)
Niina Lehtinen	Paroc Oy Ab R&D
Mark Mantonya	Owens Corning
Bhavesh Patel	Southern Research Institute
Olivier Pons Y Moll	Isover Saint-Gobain CRIR
Tim Rasinski	National Voluntary Laboratory Accreditation Program (NVLAP)
Erik Rasmussen	Rockwool International
Mitch Rose	Netzsch Instruments, North America, LLC
Roland Schreiner	FIW München
Stephen Smith	SES Consulting Services
Marc Thermitus	Consultant
Bill Thomas	Virginia Polytechnic Institute and State University (Virginia Tech)
Blaza Toman	NIST Statistical Engineering Division
Frank Tyler	Owens Corning
Robert Watters	NIST Measurement Services Division
Tom Whitaker	Industrial Insulation Group, LLC
Robert Zarr	NIST Energy and Environment Division

Fig. A1 – Attendees at 2nd Operators Workshop
on High-Temperature Guarded-Hot-Plate and Pipe Measurements

Appendix B: 1st Workshop Proceedings

	C177 (Plate)		C335 (Pipe)
1	C177 Annex A1 addresses thickness issues. It is recommended to measure thickness before & after testing. For rigid materials, contact resistance can be an issue, in addition to thickness change with temperature.	1	C335 method states that the insulation be installed, the outer circumference measured, inner circumference is assumed to be the outside of the pipe, and the thickness is calculated
2	For an ILS, it is recommended that a semi-rigid or compressible material is used. Each lab should measure thickness but test at assigned thickness. Spacers should be provided for stack designs.	2	Use wider Pi tape (2") to improve measurement of semi-rigid material
3	For rigid material, thickness should be measured insitu at start and at max temp, or pre-determined at max temp	3	Measure pipe with Pi tape to get actual pipe thickness
4	Labs can check their systems with traceable standards, at temperature, if possible	4	Use paper tape to wrap around, poke hole through overlap and then measure distance between the holes
5	Measure thickness (of semi-rigid material) at 5 locations using a calibrated vernier caliper & assign average value to specimen	5	Possibly calculate pipe dimension at temperature, @ 650C, pipe expands ~10 mm in length
6	How do you verify that plates are parallel and that this is not changing either at max temp or with time?	6	Band straps could compress material resulting in varying insulation thickness
7	Have you evaluated sample thickness measurement reproducibility & repeatability for your lab?	7	Use minimum number of bands, preferably, no more than one on test area
8	How is the test area determined?	8	Pin gauge material at temp at the top of the pipe to determine slump after the test is completed
1	How is 'uniformity of plate temperature' defined and measured? What type of thermometry & location?	1	What type of thermometry is used to measure temperature. Typical response were grounded end TC in a SS sheath. Other suggestions are ungrounded TC in metal sheath.
2	How do you address potential contact resistance issues with rigid test specimens?	2	How do you assure good thermal contact with the metal pipe? At end of groove, TC end cemented in groove touching pipe for grounding
3	Temperature measurement locations & method of measurement should be reported, sensor raw data should be reported	3	Anchor TC along pipe for at least 150mm from bead. Applies to thermocouples mounted in the pipe and on the outside surface of the test sample
4	How do you check for thermal or other degradation of sensors?	4	Bead mounted with masking tape or other tape that has emittance similar to the test material.
5	Do you use control charting [at elevated temperatures] with dedicated laboratory reference material? What material do you use? How is it verified / tracable? How do you know the control reference material is not changing?	5	Many wrap TCs around pipe test sample to maintain good contact
		6	Sometimes double layer to handle metal pipe expansions and evaluate samples at a higher mean temperature
		7	Rotate pipe to determine if TC location makes a difference
		8	Grounding helps eliminate 60 Hz hum if the data acquisition system does not have filtering

	C177 (Plate)			C335 (Pipe)
			9	Extra insulation over guards was added to reduce the power going to the guards resulting in lower TC noise.
C	**Guard Imbalance Check**		**C**	**Guard Imbalance Check**
1	What is the difference between Thermopile and TC averages for guard balancing? I.e. when thermopile is control, what are surface temperature differences?		1	How significant is the choice of using an interior thermopile versus surface mounted thermocouples to control the guard?
2	Impact of number & location of thermopile junctions		2	If TC difference is 0, thermopiles are not zero
3	Impact of number & location of TCs		3	Should gaps be insulated? (NPL)
4	Should gaps be insulated? (NPL)		4	Joints opening between the sample over the metered section and the guard section could affect results
			5	How much care is taken to assure joints are closed as pipe expands
D	**Control System Considerations**		**D**	**Control System Considerations**
1	How is the temperature controlled? If the control uses PID control, how do you assure accurate power readings (see 3)		1	How is the temperature controlled? If the control uses PID control, how do you assure accurate power readings.
2	Do you have different PID settings as the pipe temperature is changed.		2	Do you have different PID settings as the pipe temperature is changed.
3	How are you measuring power and what are sources of error?		3	C335 allows use of TC or thermopile for control
4	If you 'lock' the power, does the system drift?		4	Heater concentricity is very important
			5	What is the resolution of the data acquisition system, especially when measuring guard balance
			6	How are voltage and amps measured and how do you assure a constant standard
			7	NIST keeps shunt in oil bath
			8	Voltage Taps at test area to eliminate lead resistance
E	**Steady State Issues**		**E**	**Steady State Issues**
1	In some instances, the test meets the steady state conditions per the standard but the temperature is still changing		1	Steady state could be different (sec 8.4 in C335)
2	Check equilibrium criteria, especially in automated systems		2	The system software may detect "steady state", but is there still a longer term drift?
3	8.8.1.3 and Note 19 gives criteria using system time constant		3	What is the sampling period? What is the time constant of the apparatus?
4	What are variations or oscillations of system during control? How is 'in-control' defined & reported?		4	Does PID contol cause any fluctuations?
5	Delta T should not change by 0.2%		5	how is the ambient controlled and does that have an effect on the measurements?
F	**Surface(s) Emissivity**		**F**	**Surface(s) Emissivity**
1	Where is it important and how is it validated?		1	Check emittance of pipe
2	We need to make sure we are using a consistent procedure and reporting what the measured emissivity is.		2	Could be really important if a gap between insulation and pipe exists
4	NPL uses HE23 (Rolls-Royce paint)			
5	NIST has a ceramic coating that they are using			

21

	C177 (Plate)		C335 (Pipe)
6	C177 6.12.5 gives a method for checking emissivity but it is difficult		
1	GUM - "Guide to the Expression of Uncertainty in Measurement" (NIST Technical Note 1297)	1	Thickness has largest single impact
2	http://physics.nist.gov/pubs/guidelines/TN1297/tn1297s.pdf	2	RR density variation is probably due to thickness measurement
3	Research / provide data for random / systematic components of Q, dT, Thickness, Test Area	3	Research effect of pipe diameter; most use 3-inch pipe
		4	Possibly use wider Pi tape to help reduce variability on semi-rigid materials
		5	Should we be taking into account pipe diameter thermal expansion?
		6	Measurements influenced by fit & joint openings
		7	Measurements effect by environment and must be closely controlled
1	Choice of material? A semi-rigid material is recommended (see thickness issues)	1	Choice of material? Mineral Fiber pipe has been chosen
2	Definition of the Goal of the RR is important. Goal for next RR is to help identify apparatus variability	2	Definition of the Goal of the RR is important. Goal for next RR is to help identify apparatus variability
3	The inter-laboratory measurement & reporting protocol should be cross-checked with the findings of this workshop in order to ensure that all key target data is captured	3	Contact previous participants and try to identify more participants
4	ASTM ILS will be contacted once critical parameters are defined which are inclusive of participant equipment characteristics, and materials & temperatures established.	4	TG for review questionair and agree to specific method
5	Temperature measurement locations & method of measurement should be reported, sensor raw data should be reported	5	limit the RR max temperature to allow more participation and improve quality of data.
6	Consider sharing heater designs for possible guide to best practices	6	ASTM ILS group has agreed to do the analysis and prepare a P&B statement
7	Type of control (thermopile vs temp averaging) should be reported, or both should be used if user's system permits.	7	Proficiency testing to help monitor performance. Can we get enough participation to establish a proficiency program
8	Ruggedness testing should be done by each lab	8	Ruggedness testing should be done by each lab

2nd Operators Workshop on High-Temperature Guarded-Hot-Plate and Pipe Measurements

ASTM C16:30 Thermal Measurements Subcommittee
Test Methods C 177 and C 335

March 19-20, 2012

WELCOME & INTRODUCTIONS

William Healy

National Institute of Standards and Technology
100 Bureau Drive
Gaithersburg, Maryland, United States

NIST
National Institute of
Standards and Technology
U.S. Department of Commerce

1

Safety

Emergency Number (in-house phone): x2222

- **Fire exits**
  - Alarm notification by loudspeaker
  - Stairwells located at front, rear, and sides of building
  - Assemble at parking lot in front of Building 224
- **Shelter in place** – severe weather
- **Hazard signage (door signs)**

General Hazard Sign
(Green)

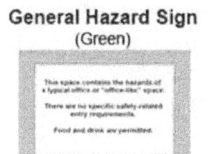

General Hazard Sign
(Red)

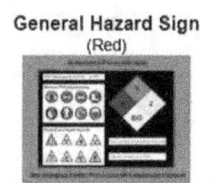

2

Announcements

DC Cherry Blossom Festival
March 20 – April 27, 2012

2nd Operators Workshop on High-Temperature Guarded-Hot-Plate and Pipe Measurements

ASTM C16:30 Thermal Measurements Subcommittee
Test Methods C 177 and C 335

March 19-20, 2012

WORKSHOP OVERVIEW

Organizers
Frank Tyler, Tom Whitaker, Robert Zarr

1

Technical Objective, Goals, and Outcomes

- *Objective:*

  To improve our understanding of the operation of the guarded-hot-plate and pipe apparatus at elevated temperatures (up to 650°C)

- *Goals:*
  - Examine, discuss, and evaluate:
    - role of NIST (reference materials and/or measurement services)
    - metrology for input (and secondary) quantities
    - control strategies for apparatus
    - sources of measurement uncertainty → GUM
  - Reduce variation in inter-laboratory comparisons

- *Outcomes:*
  - Notebook (copies of slide presentations, 1st Workshop summary)
  - Addendum (written summary of Discussion edited by Workshop Organizers and reviewed by WERB for distribution 05/31/2012)
  - Action items will be referred to the appropriate ASTM task group

2

Background and Approach

- *History*
  - 2007 – 1st Operators Workshop, Owens Corning, Granville
  - 2012 – 2nd Operators Workshop, NIST Gaithersburg

- *Background:*
  - 1st Workshop:
    - ➢ Presented plate and pipe methods separately
    - ➢ Summary report, however, noted commonality of methods

- *Current Approach:*
  - 2nd Workshop builds on the results of the 1st Workshop
  - Technical sessions presented generically taking advantage of common issues

3

Comparison of Plate and Pipe Methods

	Plate	Pipe
Equations (C 1045)	$\lambda_a = \dfrac{QL}{A(T_h - T_c)}$	$\lambda_a = \dfrac{Q \ln\left(\dfrac{r_{out}}{r_{in}}\right)}{2\pi L_p (T_{in} - T_{out})}$
Differences	axial heat flow (shape factor)	radial heat flow (shape factor)
Similarities	power (DC voltage and resistance), temperature, thermopile gap measurements, (PID) control, GUM, inter-laboratory design	

4

Questions of Interest

- *What is the role of NIST in providing reference materials or measurement services?*
    - ASTM C16:30.1.2 Standard Reference Materials Task Group

- *What are the benefits of metrological traceability for laboratory measurement process?*

- *How to model apparatus for PID simulation control?*

- *What is the GUM? How do the GUM and inter-laboratory comparisons complement each other?*
    - Evaluation of measurement uncertainty and precision

- *How does one design an inter-laboratory comparison?*
    - Identification of the goals, test conditions, etc.

5

Scope

- *Topics covered in 4 Sessions*
    1. High-temperature reference materials/measurement services
    2. Measurement metrology
        - Electrical and dimensional tours
        - Guarded hot plate presentation and tour
        - Input and secondary quantities
            - Thickness
            - Sensor reliability (temperature, power)
            - Guard imbalance
            - Surface emissivity
    3. (PID) Control
    4. Uncertainty (GUM) and Inter-laboratory comparisons

6

Guidelines and Announcements

- *Please turn off cell phones, beepers, pagers, etc.*

- *Session format*
  - *Speakers*: 15 minute presentation; 5 minute Q/A
  - *Audience*: **hold** questions until Q/A time slot (additional Q/A time during Break or Discussion)

- *Amenities*

- *Monday night dinner* (optional)
  - Local restaurant
  - Reservations to Monyelle by 12 noon

7

Questions?

8

2nd Operators Workshop on High-Temperature Guarded-Hot-Plate and Pipe Measurements

High-Temperature Thermal Insulation
Industry Needs

Thomas Whitaker
Chairman, ASTM Committee C16 on Thermal
Insulation

1

ASTM Committee C16
Thermal Insulation

- ASTM Committee C16 on Thermal Insulation was formed in 1938. C16 meets twice a year, in April and October, with about 120 members attending over three days of technical meetings capped by a discussion on relevant topics in the Thermal Insulation industry. The Committee, with a membership of approximately 350, currently has jurisdiction of over 145 standards, published in the Annual Book of ASTM Standards, Volume 04.06. C16 has 8 technical subcommittees that maintain jurisdiction over these standards.

- These standards have and continue to play a preeminent role in all aspects important to the industry of thermal insulation, including products, systems, and associated coatings and coverings, excluding refractories.

- The scope of the Committee is the development of standards, promotion of knowledge, and stimulation of research pertaining to thermal insulation materials, products, systems, and associated coatings and coverings.

- These activities are coordinated with those of other ASTM Committees and national and international organizations having similar interest.

2

Industrial Manufacturing

- Energy Use in the Manufacturing Sector in the United States
- To predict Usage and/or potential savings, the industry needs reliable thermal properties of high temperature insulation materials
- How reliable is our data and How can we improve it

Energy Use in the Industrial Sector

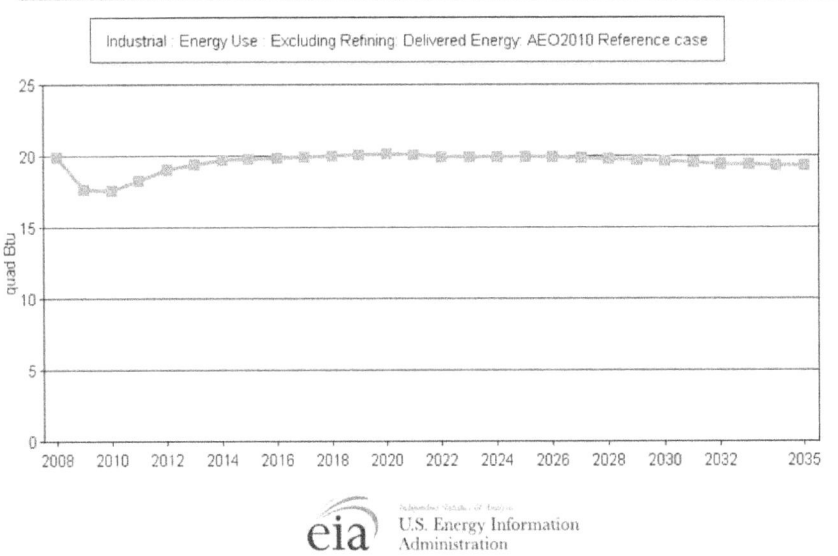

Industrial : Energy Use : Excluding Refining: Delivered Energy: AEO2010 Reference case

US Government Commitment

DOE Awards More Than $30 Million to Help Universities Train the Next Generation of Industrial Energy Efficiency Experts

September 13, 2011
WASHINGTON, D.C. – U.S. Energy Secretary Steven Chu today announced more than $30 million for 24 universities in 23 states across the country to train undergraduate- and graduate-level engineering students in manufacturing efficiency to help them become the nation's next generation of industrial energy efficiency experts. Each school will receive $200,000 to $300,000 per year for up to 5 years to help university teams to gain practical training on core energy management concepts through DOE's successful Industrial Assessment Center program.

5

Energy Efficiency Experts, Design Engineers, Reliability Engineers

How reliable are the Thermal Properties of High-Temperature Insulation Materials?

What is the impact of the quality of the Thermal Properties?

6

ASTM C335 – Pipe Apparatus

Tests performed at seven different laboratories using the horizontal guarded-end apparatus and at one laboratory using an unguarded cylindrical screen test apparatus on two samples of calcium silicate insulation in the range of mean temperatures from 35 to 390°C did not vary by more than 6.3 % of the average

ASTM C177 Guarded Hot Plate

In 1988, results of a interlaboratory comparison were reported for seven high-temperature guarded-hot-plate apparatus. The plates ranged in size from 203 to 406 mm in diameter and 300 to 610 mm². Different matched pairs of fibrous alumina-silica and calcium silicate were measured by each laboratory over a mean temperature range from 330 to 701 °K. Reference equations based on NIST-Boulder corrections were fit to the data. Imprecision in the deviations from the model were 15 and 16 % (2s level) for fibrous alumina-silica and calcium silicate, respectively. It was established that a significant percentage of the standard deviation in this comparison was due to material variability and not apparatus error.

A Round Robin Interlaboratory Comparison of Thermal Conductivity Testing Using the Guarded Hot Plate up to 1000°C

- Published January 2002, ASTM STP 1426EB

- A round robin interlaboratory comparison of thermal conductivity measurements was performed using the Guarded Hot Plate apparatus at temperatures from about 0°C to 1000 °C. There were twelve participating laboratories including four national laboratories in three different countries. A statistical analysis was performed and the variation in measurement results is discussed. Both within laboratory and between laboratory variability is analyzed. A definite conclusion of this comparison is that measurement variation increases progressively with increasing temperature. As a result there is a need for the U.S. national laboratory, the National Institute of Standards & Technology, to develop high temperature testing capability and then high temperature thermal conductivity reference standards.

8

How Can We Improve the Data?

- Improve the Test Methods
- Improve the Apparatus
- How does an operator know if they are providing accurate information?
  - There are no SRM materials for comparison
  - There are no Proficiency Programs for high-temperature thermal conductivity measurements, either for Plate or Pipe.
  - Any commercial or corporate laboratory cannot demonstrate proficiency nor demonstrate their high-temperature measurements are in agreement with measurements made by NIST or other national metrology institute.

10

Impact of the Reliability of High Temperature Thermal Properties

- Safety, Personnel Protection, Fire Protection
- Some Processes require critical control
- Control Energy costs
- Control Greenhouse Gas Emissions by reducing waste heat loss

11

What Does the Industry Need?

- ASTM needs to hear from you regarding changes to the method and/or apparatus;
  - reduce variability
  - improve the accuracy & repeatability of measurements
- Participation from NIST to provide an SRM and provide Transfer Standards that can be used to demonstrate traceability.
- Encourage NIST to participate in comparisons with other National Laboratories
- Encourage research projects into measuring and verifying the quality of high-temperature thermal conductivity measurements.

12

Questions & Discussion

13

35

European situation on high temperature testing

Erik Rasmussen
Public Affairs
Rockwool International A/S

Legal background

- EN standards for Thermal insulation products for building equipment and industrial installations have been published in December 2009 and CE marking will be mandatory as of August 1, 2012
- Thermal conductivity shall be declared for the temperature range for which the products can be used.
- Testing shall be performed by an accredited and notified laboratory for the initial type testing

Voluntary initiatives

- The current VDI certification is expanded to a VDI & Keymark cooperation for a joint European product certification with focus on thermal conductivity accuracy in testing (both for manufacturers and 3rd parties)
- Roland Schreiner will present the work of FIW on defining a material to be used for testing as no reference material is available
- Helge Høyer will later present the results of a recent round-robin on GHP on the same one set of test specimens.

ROCKWOOL
FIRESAFE INSULATION

Expanded glass granulate
as a Round Robin material for
thermal conductivity to higher temperatures

Dipl. – Ing. Roland Schreiner

ASTM Workshop 2012, NIST

Qualification of the Round Robin material

- temperature resistance up to 500 °C
- repeatability of the temperature exposure
- availability in different thicknesses
- suited for different measuring methods
- uniform apparent density in an acceptable range
  - incompressible
- uniformity of the test material and the test specimen.
  - homogeneity
  - isotropy.
- Heat transfer model of „true thermal conductivity" possible
  - Temperature dependent
  - Include all heat transfer mechanism e.g. radiation

ASTM Workshop 2012, NIST

Levels of Round Robins

1. Round Robin with **one test material** and **one test method** but different apparatus designs
Comparisons and outliers only in relation to a mean value

2. Round Robins with **one test material** and **different test methods** and different apparatus designs
Comparisons and outliers in relation to a „better" mean value

3. **Heat transfer model calculation** of the test material to evaluate expected values. **Level 1** needed to adjust material properties (spectral parameters and apparent extinction)
Comparison and outliers in relation to the expected temperature dependent *„true thermal conductivity"*

Properties of the expanded glass ganulate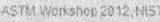

Expanded glass granulate
Bead size: 1 – 2 mm
Bulk Density: 250 kg/m³ ± 10%
Thermal conductivity level at 10 °C approx. 0,070 W/(m·K)
Maximum of operational temperature = 550 °C

39

The four pillar policy

**1. Guarded
Hot Plate**
20 – 100 mm
in a CaSi-Frame

2. Pipe Tester
Diameter 20 – 324 mm
With plexiglass tubes or metal cladding

3. Nusselt Sphere (small and big)

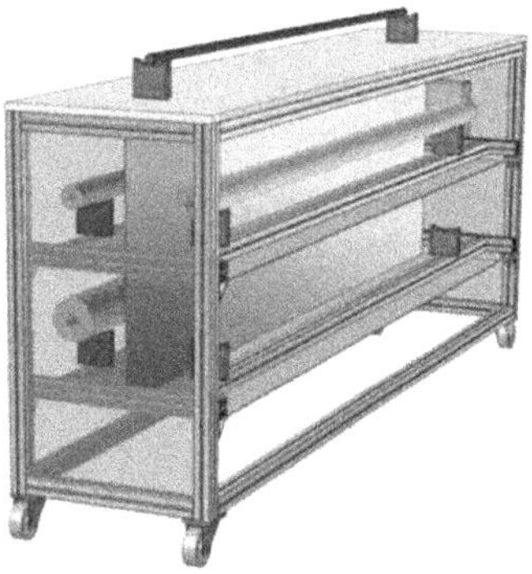

4. Heat transfer model of porous media

ASTM Workshop 2012, NIST

5

The long way to the „true thermal conductivity"

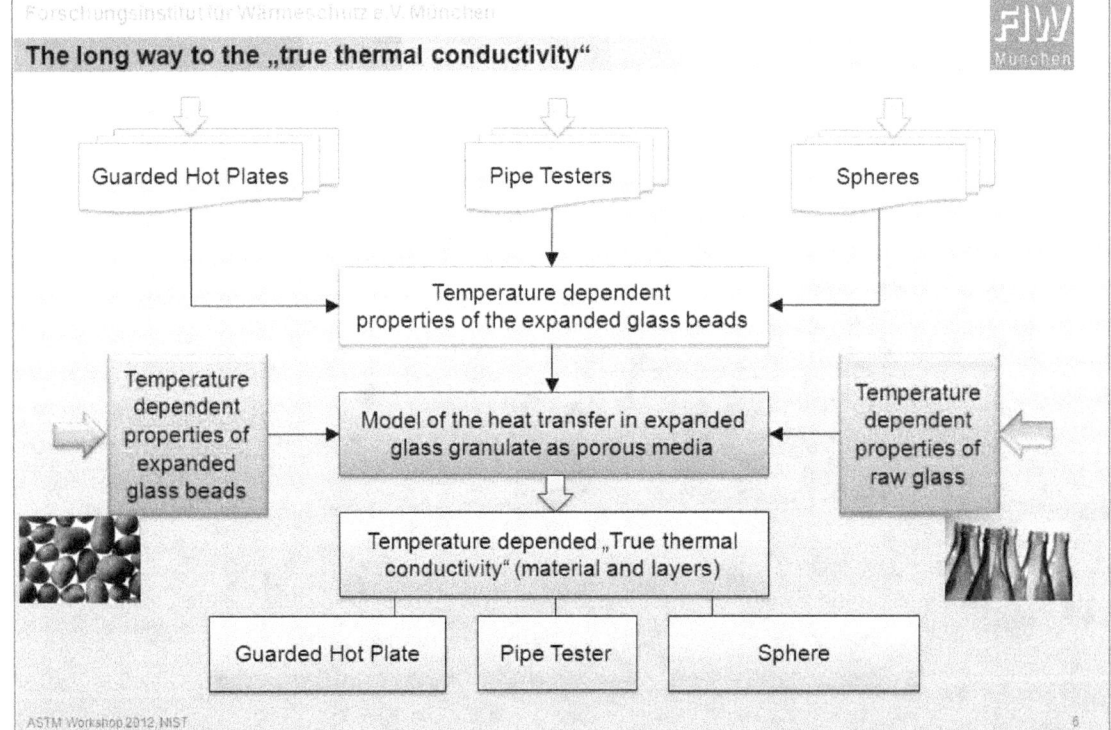

ASTM Workshop 2012, NIST

6

40

The European Glass beads Round Robin

Results of 5 laboratories

Method	Round Robin Level 1 one test material, one test method			Round Robin Level 2 one test material, different test methods			Round Robin Level 3 True Thermal Conductivity with HTM
	compare with	number	standard deviation W/(m·K)	compare with	number	standard deviation W/(m·K)	
Sphere	mean value	6	-	recalculation of single values			Integral value of Thermal transmissivity λ_1
Pipe Tester	mean value	65	0,002	Mean value of thermal transmissivity	201	?	crossover function
GHP	mean value	130	0,003		λ_{GHP} + apparent extinction		Transfer factor Layer 50/100 mm

Tests of glass beads and raw material   Statistic correlation between spectral parameters and apparent extinction

Mean value = polynomial regression 3rd degree
HTM = Heat transfer model of expanded glass granulate
Crossover function =

$$\lambda_{GHP} = \cfrac{1}{\cfrac{E^*}{4\sigma T^3} + \cfrac{1}{\lambda_1}}$$

completed activities

future activites

The Transfer factor (GHP) – Thermal transmissivity integral (Pipe Tester)

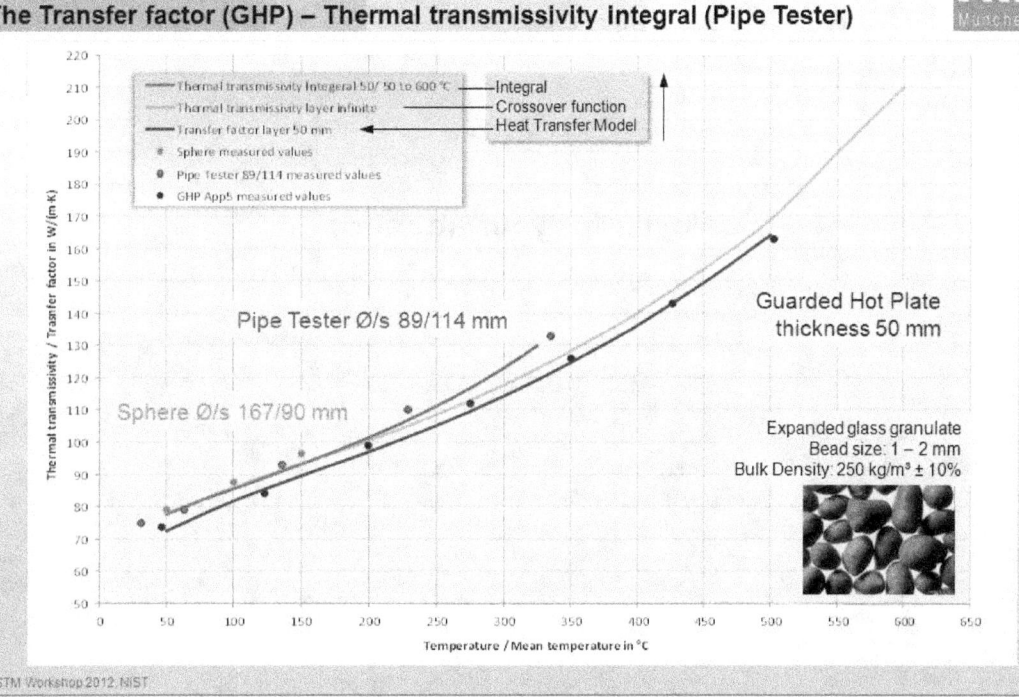

41

Future perspectives

- Improvement of measured values by means of the Gaussian correction to evaluate the uncertainty of the reference curve of ± 3% (now only 9% of all measured values exceed the ± 3% limit)

- Add more values to the Round Robin (Labs and apparatus)

- Improve the heat transfer model and the properties of the expanded glass beads with the focus on radiation and uncertainty budgets

- Bring the Round Robin to 3. level stage

- Extend the temperature range to the very low temperatures (-180 °C)

- Start of a new Round Robin this year ("AKT Thermophysik")

- Discussion of *true thermal conductivity* to higher/lower temperatures

9

Thank you for your attention!

10

42

NIST Guarded-Hot-Plate Facilities

Thermal Insulation Standard Reference Materials (SRMs) and Measurement Services

March 19, 2012

Robert Zarr and William Healy

National Institute of Standards and Technology
100 Bureau Drive
Gaithersburg, Maryland, United States

National Institute of
Standards and Technology
U.S. Department of Commerce

2nd Operators Workshop on High-Temperature
Guarded-Hot-Plate and Pipe Measurements

1

Interactions Between ASTM C16.30 and NIST

- **ASTM Documentary Standards**
  - C 177 (Dickinson, Robinson, Powell, Hust, Smith)
  - C 1043 and C 1044 (Rennex, Flynn, Hahn, Peavy, Zarr)
  - C 1558 (Dalton, Zarr)

- **NIST Thermal Insulation Standard Reference Materials (SRMs)**
  - *ASTM STP 660* – "position" paper advocated an SRM approach (1977)
    - ➢ SRM 1450, Fibrous Glass Board
    - ➢ SRM 1451, Fibrous Glass Blanket
    - ➢ SRM 1449, 1459, Fumed Silica Board
  - ASTM C16:30.1.2 SRM Task Group (2005, H. Hoyer, Chair)

- **NIST High-temperature Guarded-Hot-Plate Laboratory Facility**
  - C16:30 letter (D. McCaa) to A. Prabhakar, NIST Director, requested relocation of high-temperature GHP apparatus from Boulder to Gaithersburg (1994)
  - NIST Director A. Prabhakar responded affirmatively citing industry needs; Boulder GHP apparatus assigned to the Heat Transfer Group

2nd Operators Workshop on High-Temperature
Guarded-Hot-Plate and Pipe Measurements

2

What is the role of NIST in providing thermal insulation reference materials or measurement services?

100 years GHP technology (1912-2012)

NIST 1016 mm GHP apparatus (line-heat-source → ASTM C1043)

NIST 200 mm GHP apparatus (left) Powell, Watson, Siu (seated)

(→ ASTM C 177-45)

NIST 500 mm GHP apparatus

2nd Operators Workshop on High-Temperature Guarded-Hot-Plate and Pipe Measurements

3

Recent Technical Activities
NIST 1016 mm Guarded-Hot-Plate Facility

- **NIST Calibration Service for Thermal Resistance (2009)**
  - Service ID Numbers: *36110C to 36199S*
  - Offered annually (October 1) for U.S. customers
  - T_m of 24 °C (75 °F); $U(R)$ = 1 % to 3 % (k = 2)
  - http://www.nist.gov/calibrations/thermal_resistance.cfm

- **Uncertainty Publications (GUM)**
  - *J. Res. Natl. Inst. Stand. Technol.* **115**, 23-59 (2010)
  - *ASTM JOTE*, **38**, *No.2* (March 2010)

- **NIST SRM 1450d, Fibrous Glass Board (2011)**
  - *NIST Special Publication 260-173* (August 2011)
  - Thermal conductivity (λ): 280 K to 340 K
  - Unit size: 610 mm x 610 mm x 25 mm
  - http://www.nist.gov/srm/index.cfm

2nd Operators Workshop on High-Temperature Guarded-Hot-Plate and Pipe Measurements

4

NIST SRM 1450 Series Sales History

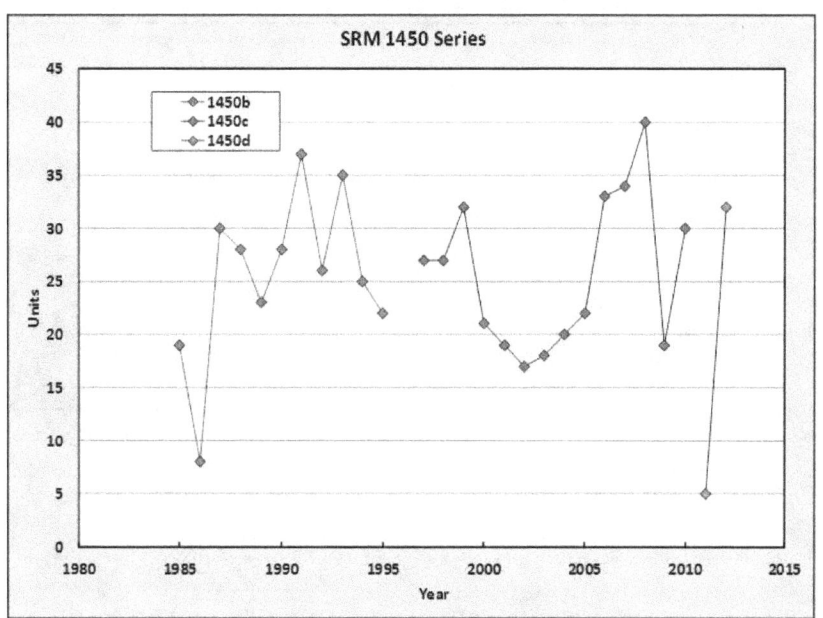

NIST Quality System

- **Motivation: Mutual Recognition Arrangement (MRA) signed by the directors of 38 national metrology institutes (1999)**

 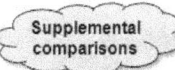 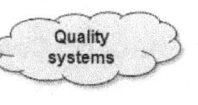

- **Major Components**
  - Multi-tiered (http://www.nist.gov/qualitysystem/):
    - ➢ QM-I (NIST-wide)
    - ➢ QM-II (division level)
    - ➢ QM-III (service level – measurements and SRMs)
  - Based on ISO/IEC 17025 and ISO/IEC Guide 34
  - Uncertainty - compliant with GUM
  - Assessment timetable: 5 year cycle
  - General schedule
    - ➢ primary units (such as kelvin, among others)
    - ➢ derived units (such as thermal conductivity)

Quality System for NIST GHP Laboratory

- **Quality System Assessment**
  - Feb. 2010: 1st assessment by NIST
  - Oct. 2010: Conformance declared with NIST Quality System
  - Nov. 2010: Approval by Inter-American Metrology System (SIM) Quality System Task Force (regional metrology organization for the Americas under the BIPM)

International Developments

- **BIPM – created in 1875**
  (Bureau International des Poids et Mesures)

- **Consultative Committee on Thermometry (CCT) – created in 1938**

- **Working Group 9 on Thermophysical Properties – created in 2001**
  - 2012 NMI membership: NMIJ/AIST, NPL, LNE, I.N.RI.M, PTB, NIST, KRISS, VNIIM, CENAM, NIM
  - Current activities: pilot inter-laboratory comparisons
    - Thermal conductivity by GHP method: LNE (pilot)
    - Laser-flash thermal diffusivity: NMIJ (pilot)
    - Normal spectral emissivity: NIST (pilot)

NIST High-Temperature GHP

- **1996 to 2000 – Evaluation of previous GHP equipment**
    - Bora Rugaiganisa, Guest Worker, evaluated 200 mm low-temperature GHP
    - Daniel Flynn, Contractor, evaluated 250 mm high-temperature GHP
        - Electrical cross-talk noted between heaters and PRTs
        - Difficulty with constant-power and constant temperature control
        - Analysis of plate temperatures and edge heat flows
        - Mechanical operation unwieldy

- **2001 to present – 500 mm GHP apparatus**
    - 2001 to 2004: specifications, design, components construction
    - 2005 to 2006: assembly of apparatus
    - 2007 to 2010: verification and PID control testing

Future Considerations

- **Inter-laboratory comparisons with other NMIs**
    - Temperature: – 20 °C to 200 °C (ΔT: 25 K and 50 K)
    - Pressure: ambient to 10^{-4} torr (fixed mean temperature and ΔT)

- **Calibration Services (U.S. customers)**
    - Individual measurements (initially at cost, several $K)
    - Temperatures: ambient to 200 °C (ΔT: 25 K and 50 K)
    - Thicknesses: 25 mm to 50 mm
    - Different material(s) for each customer

- **SRMs**
    - Benefit: one material for everyone
    - Batch certification (100 % sampling possible)

High-Temperature SRM Candidates

- **ASTM input**
  - ASTM Questionnaire distributed to C16, E06, and E37
  - Response preferences:
    - ➢ 300 mm square, 25 mm thick, board form
    - ➢ Maximum temperature: 650 °C (mean or face)
  - Working with C16:30.1.2 to identify potential candidate materials

- **Potential candidates (in no specific order)**
  - Mineral wool board
  - Unbonded (needled glass-fiber) blanket
  - Calcium silicate board
  - Fumed silica board (SRM 1449, 1459)

- **Assessment**
  - HFM tests (24 °C) before and after conditioning in air at 650 °C
  - Thermogravimetry analysis (TGA)

Comments and Questions

- **General plan** (interested in your input for the discussion session)

> **Immediate**
> NMI inter-laboratory
> comparisons

> **Short-term**
> Measurement services
> for public

> **Long-term**
> High-temperature
> SRMs

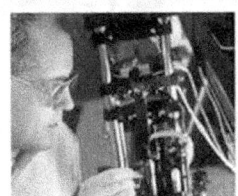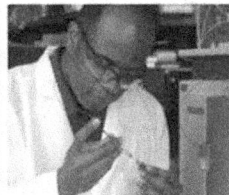

NIST Measurement Services

Robert L. Watters, Jr.
Associate Director for Measurement Services

National Institute of
Standards and Technology
U.S. Department of Commerce

**MATERIAL
MEASUREMENT
LABORATORY**

NIST Measurement Services

- Performed by NIST
  - Publications on measurement science research
  - Fee-supported services
    - Calibration services
    - Standard Reference Data
    - Standard Reference Materials
    - Laboratory accreditation services (NVLAP)
- Resources for Customers
  - Services for legal metrology labs
  - Metrology training
  - Measurement practice guides
  - User facilities (CNST and NCNR)

NIST   MATERIAL MEASUREMENT LABORATORY   2

Fee-Supported Services

- Calibrations
  - Service in NIST technical lab
  - Customer sends instrument to NIST
- Standard Reference Data (SRD)
  - Evaluated numeric data on physical or chemical properties
  - Scientific algorithms on behavior of systems
- Standard Reference Materials (SRMs)
  - Physical artifacts with certified physical or chemical properties
- Laboratory Accreditation
  - Formal assessment of quality systems

NIST

MATERIAL MEASUREMENT LABORATORY

3

Program Output Trends

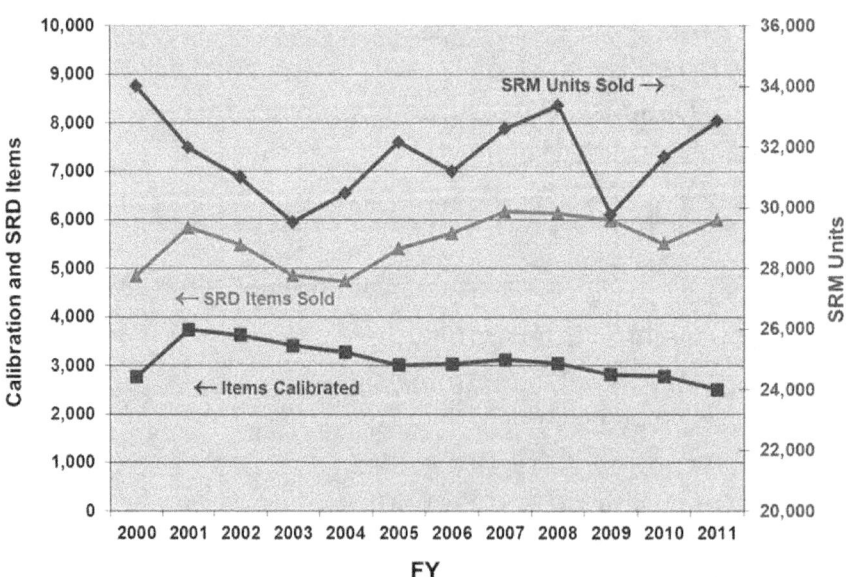

NIST

MATERIAL MEASUREMENT LABORATORY

4

Program Income Trends

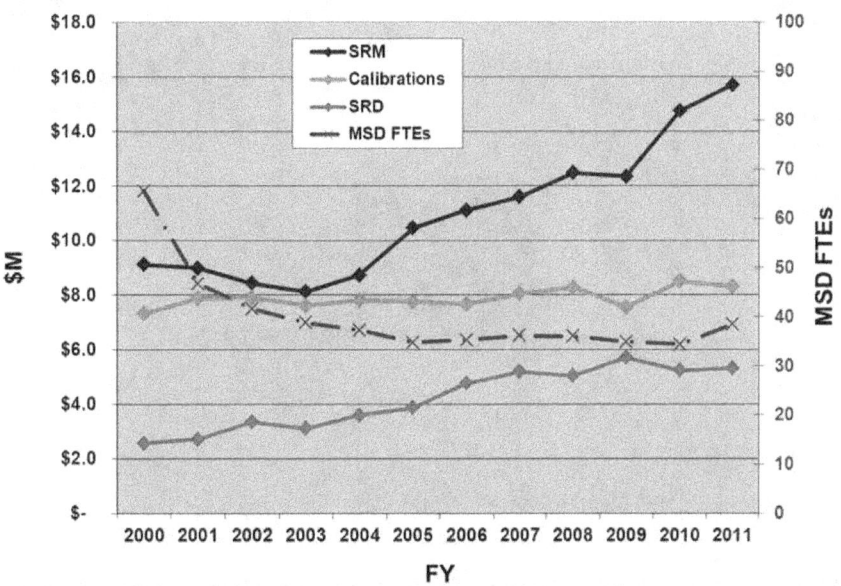

Calibration Services

- 13 Divisions in mostly in PML
- 7 major categories
  - Dimensional
  - Electromagnetic
  - Ionizing Radiation
  - Mechanical
  - Optical
  - Thermodynamic
  - Time and Frequency
- Per year:
  - 2,800 items
  - 25,000 tests
  - >600 unique customers
  - $8 M income
  - 1,800 customer transactions

Calibration Income by Metrology

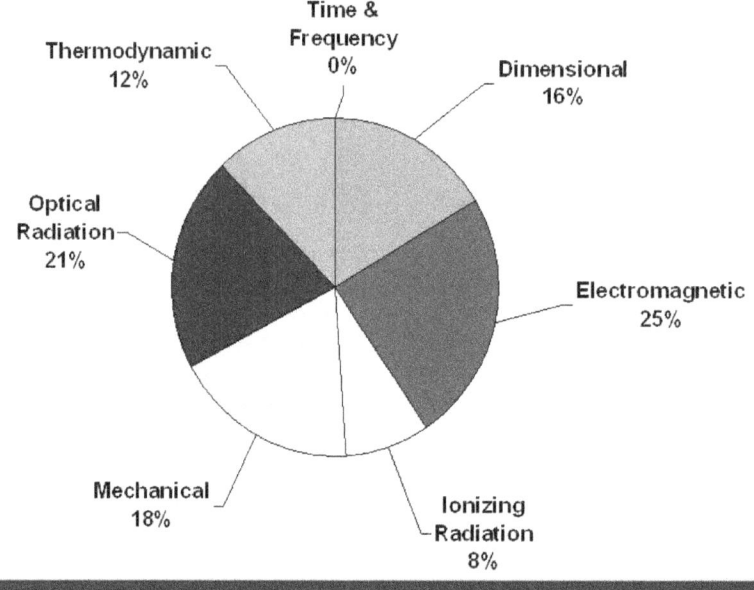

Items Calibrated by Customer Type

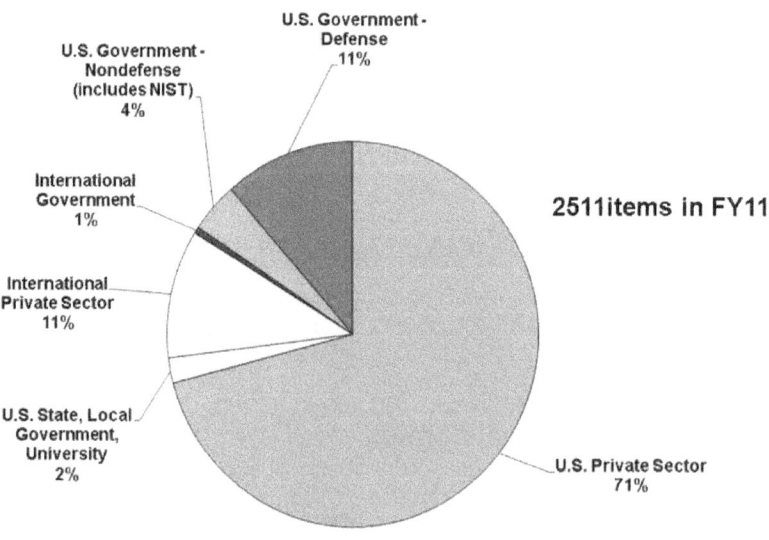

2511items in FY11

NIST Scientific Data

- All Labs; most divisions
- Standard Reference Data
  - 41 PC products available
  - 71 Online SRD systems out of 96 total NIST systems
  - 6,000 units sold/year
  - 19.1 M/year data downloads
  - $5.3 M in sales

Technical Areas for NIST SRD

- Analytical Chemistry
- Atomic and Molecular Physics
- Biotechnology
- Chemical and Crystal Structure
- Chemical Kinetics
- Environmental data
- Fire
- Fluids
- International Trade
- Law Enforcement
- Materials Properties
- Optical Character Recognition
- Surface Data
- Text and Video Retrieval
- Thermophysical & Thermochemical

MSD Sales and Customer Service for SRD Purchased Products

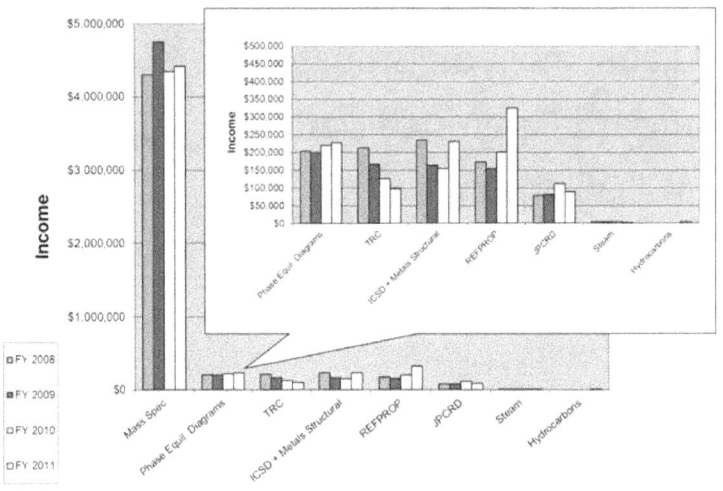

Free Online Systems – Top Downloads

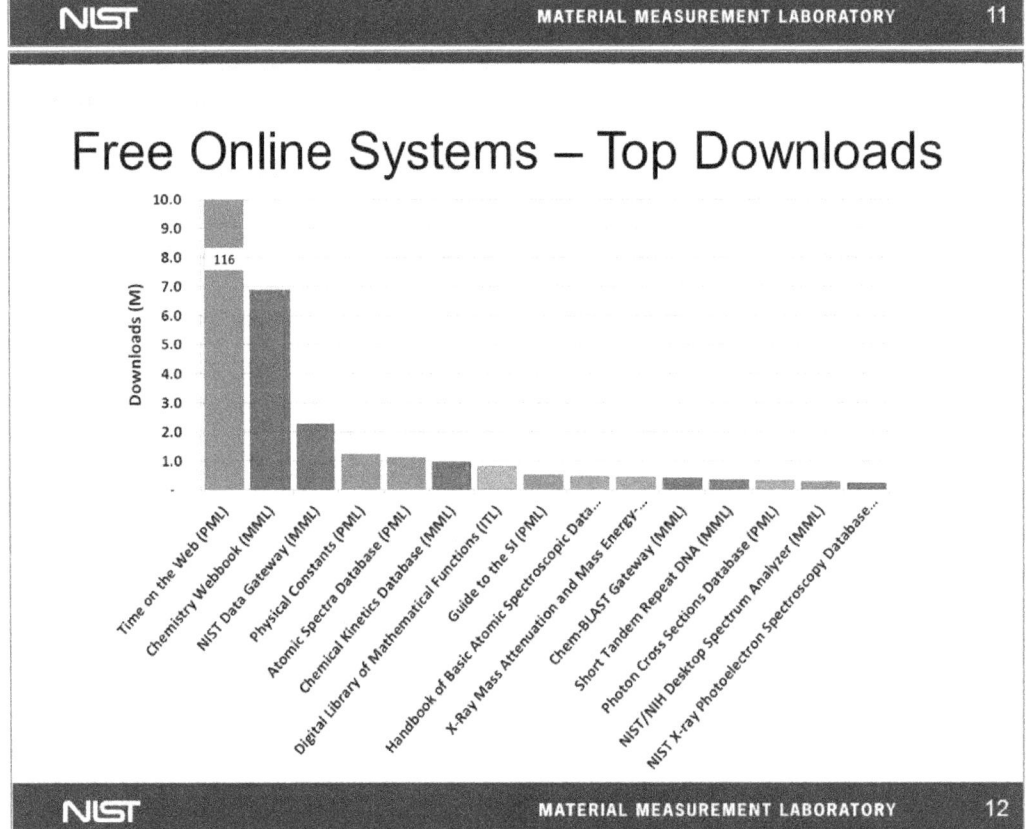

Journal of Physical and Chemical Reference Data

- NIST partner with American Institute of Physics
- Business Editor Bob Watters (MML)
- Co-editors Allan Harvey and Don Burgess (MML)
- 25 – 30 articles; 2000 pages per year
- 4000 citations per year
- Citation half-life >10 years
- High impact factor over the years
  - Average number of citations for "recent" articles

JPCRD Impact Factor

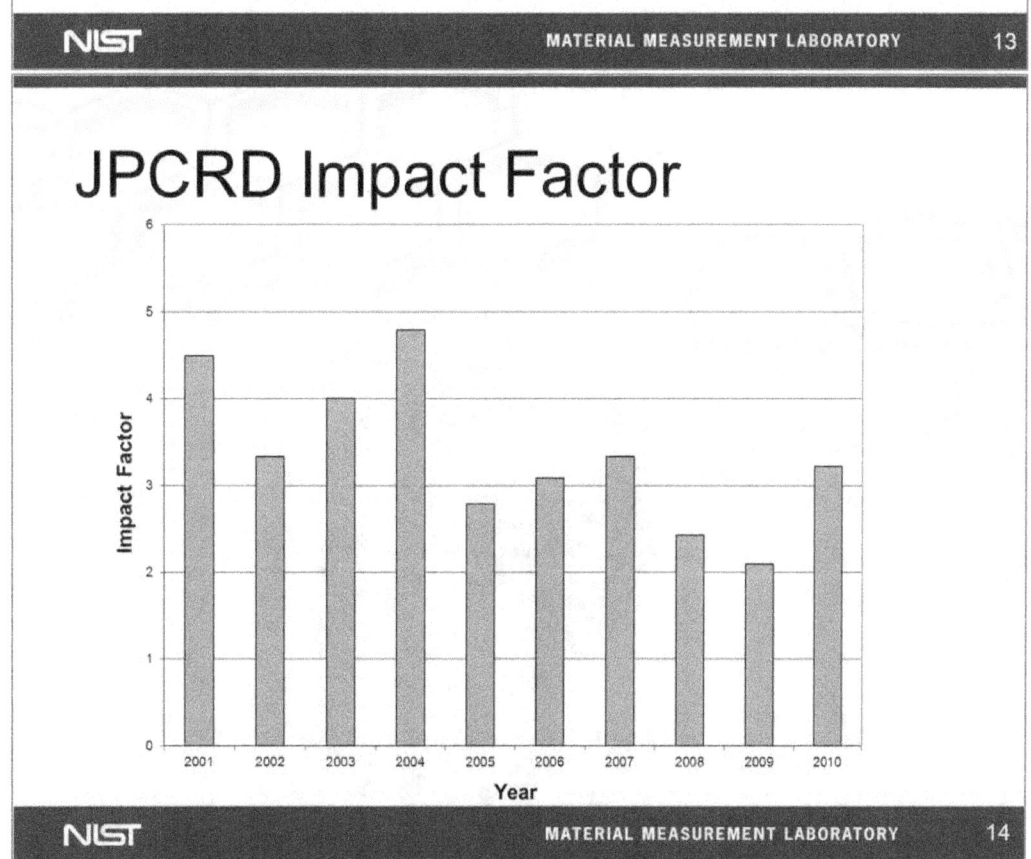

Standard Reference Materials

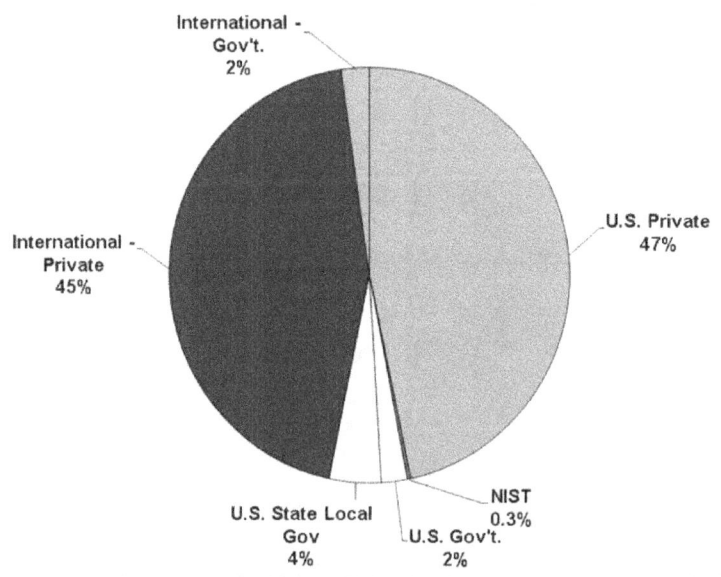

- 21 Divisions in 3 NIST
- Laboratories
- 3 major categories
  - Chemical composition, physical properties and engineering properties
- ~ 1300 products
- Approx. 32,000 units sold/year, with $15M income

SRM Units Sold by Customer Type

International -
Gov't.
2%

U.S. Private
47%

International -
Private
45%

NIST
0.3%

U.S. State Local
Gov
4%

U.S. Gov't.
2%

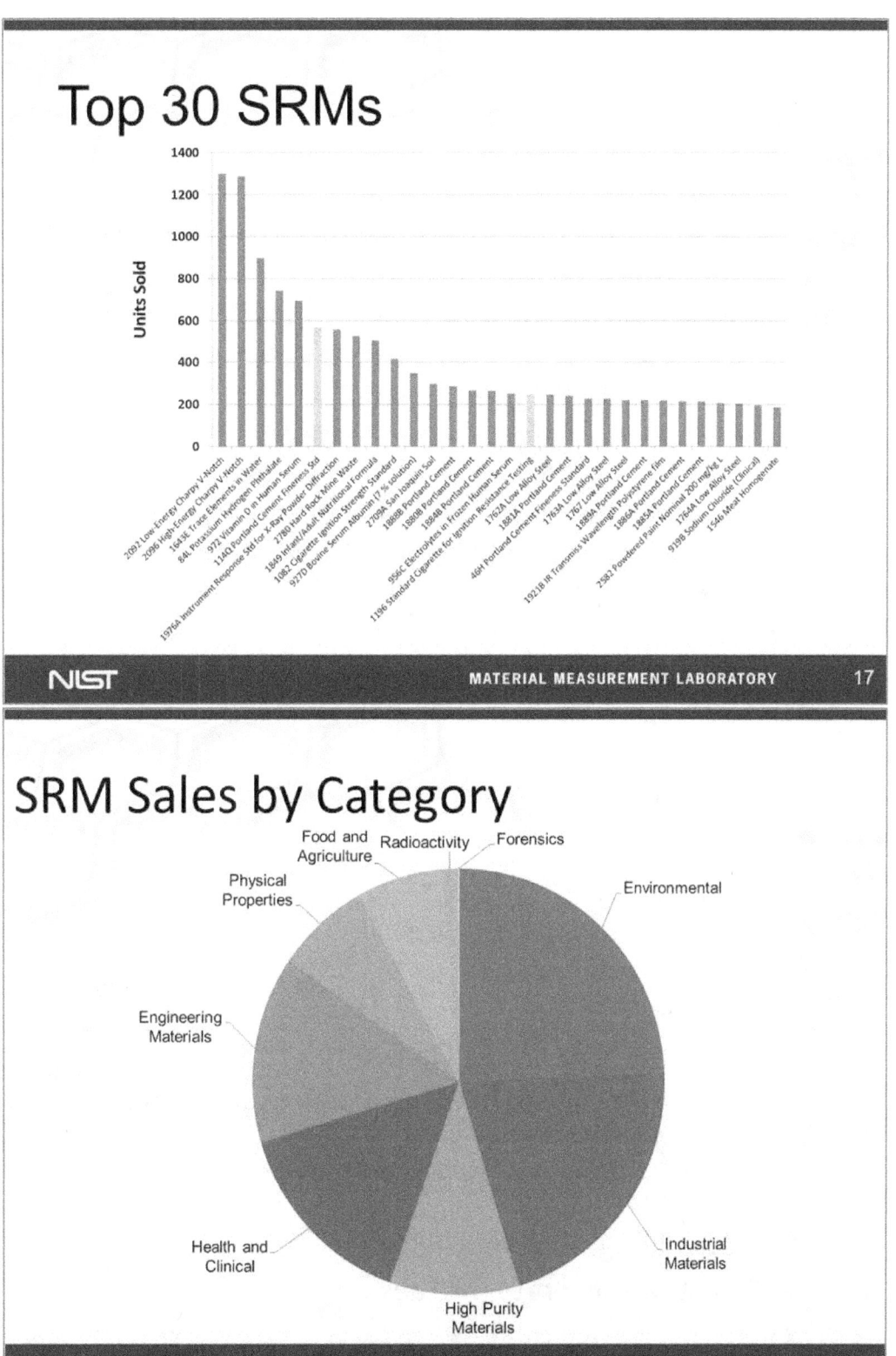

Top 30 SRMs

SRM Sales by Category

Category History

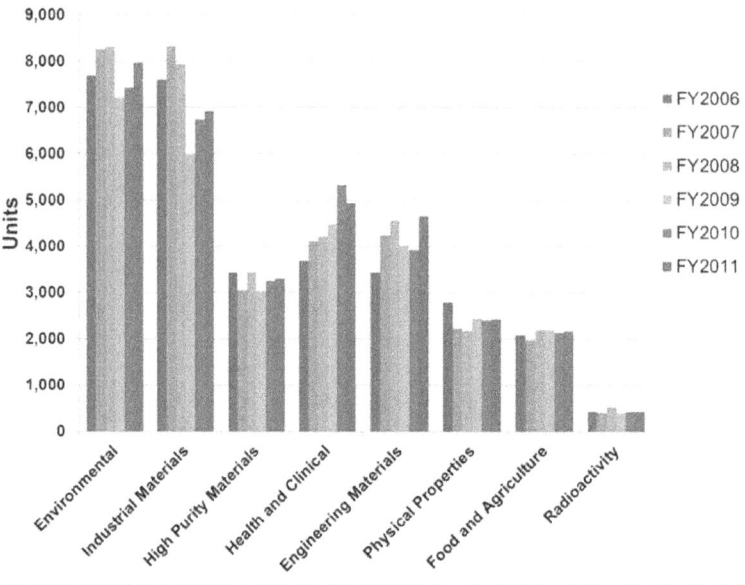

MATERIAL MEASUREMENT LABORATORY 19

Roles and Responsibilities

- Laboratories
  - Technical responsibility
    - Plan, set priorities, fund and implement the development, production, delivery, and ensure the quality of SRMs
- Measurement Services Division
  - Business, administrative, and product support services
    - Includes material preparation, inventory management, documentation support, business and administrative support
- Statistical Engineering Division
  - Statistical support to the Labs

MATERIAL MEASUREMENT LABORATORY 20

SRM – Overall Process

- Identification of stakeholder measurement needs
- Underlying research needed to address the issues
- Decision on the best channel to fulfill the needs
- SRM development work to determine feasibility
- Submission of production plans
  - Statements of Work (SOWs)
- Business case check
- Material Hazard Assessment
- Production measurements and certification

SRM Statements of Work

- Alignment with NIST priorities
  - Is the SRM the optimum measurement service solution?
  - Is the SRM consistent with each Laboratory's mission priorities?
- Justification
  - Engagement with user community
  - Drivers such as industry need for better accuracy, traceability, regulatory compliance
  - Are there other ways to fulfill the need?
    - Published measurement method
    - MAP
    - Calibration service
  - Renewals forever, or other perpetuation plans?
- Technical
  - Measurands to be value-assigned
  - What measurement design will be used?
    - All within NIST
    - NIST + outside collaborators
    - Only outside collaborators with a NIST-designed protocol and data analysis
  - SED design and analysis or existing SED-approved template

Modes Used at NIST for Value-Assignment of Reference Materials for Chemical Measurements

	C	R	I
1. Certification at NIST using Primary Method with confirmation by other methods.........	♦		
2. Certification at NIST using two independent critically evaluated methods.................	♦	♦	
3. Certification/value assignment using one method at NIST and different methods...... by outside collaborating laboratories	♦	♦	
4. Value assignment based on measurement of two or more outside collaborating........ laboratories using different methods		♦	♦
5. Value assignment based on a method dependent (procedurally-defined) technique..		♦	♦
6. Value assignment based on NIST measurements by a single method (but does............ not meet criteria for certification		♦	♦
7. Value assignment based on outside collaborating laboratory measurements............... using a single method		♦	♦
8. Value assignment based on selected data from interlaboratory studies.................		♦	♦

Key: **C = Certified Value**, R = Reference Value, I = Information Value

Perpetuation Plans

- What happens if the user community gets "hooked" on an SRM?
  - Lab committed to renewals forever?
  - Other means to achieve the same driver goals
  - Might explore a renewal certification mode that uses only outside collaborators (certification mode 4) – could be a good topic for SD support
    - High volume sales
    - Existing SRM can be used as a control
    - Methods well-established
    - Source of material similar to original SRM
    - Suitable collaborators identifiable
      - NMIs with proven performance in Key Comparisons

High- and low- temperature thermal conductivity measurements

French situation

A. Koenen

J. Hameury - B. Hay - E. Guillaume

ASTM Workshop - NIST March 2012

1

-  **CE marking in Europe**
  - ✓ Product range
  - ✓ temperature range
- **What type of measurement**
- **Inter-comparison**
  - ✓ At room temperature
  - ✓ For cryogenic application
  - ✓ For High temperature
- **Need of insulation reference materials**
- **Situation at LNE**
  - ✓ GHP, "small" high temperature GHP

LNE

ASTM Workshop - NIST March 2012

2

Existing in its present form since 1993, the CE marking is a key indicator of a product's compliance with EU legislation and enables the free movement of products within the European market

Thermal insulation products for building equipment and industrial installations

- **EN 14303** : mineral wool **(MW)**,
- **EN 14304** : elastomeric foam **(FEF)**,
- **EN 14305** : cellular glass **(CG)**,
- **EN 14306** : calcium silicate **(CS)**,
- **EN 14307** : extruded polystyrene foam **(XPS)**,
- **EN 14308** : polyurethane foam (PUR) and polyisocyanurate foam (PIR) products
- **EN 14309** : expanded polystyrene **(EPS)**,
- **EN 14313** : polyethylene foam **(PEF)**,
- **EN 14314** : phenolic foam **(PF)**,

- **NF EN ISO 13787** : Determination of declared thermal conductivity (ISO 13787:2003)

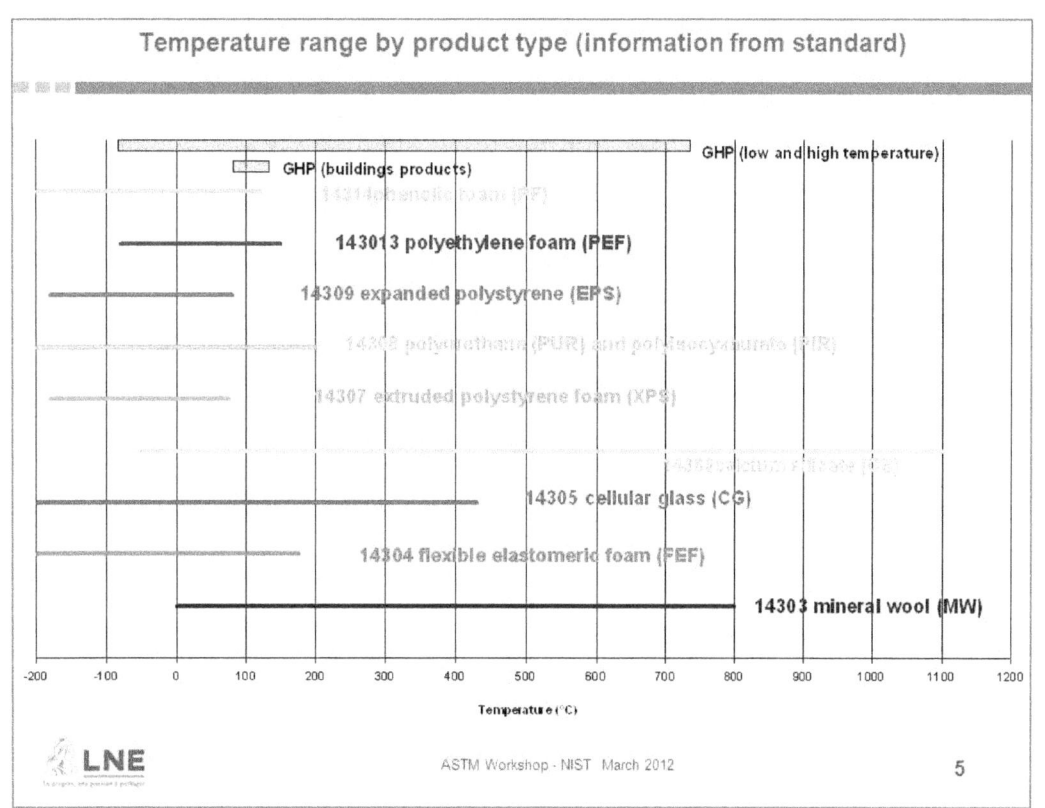

Temperature range by product type (information from standard)

GHP (buildings products)   GHP (low and high temperature)

14314 phenolic foam (PF)

143013 polyethylene foam (PEF)

14309 expanded polystyrene (EPS)

14308 polyurethane (PUR) and polyisocyanurate (PIR)

14307 extruded polystyrene foam (XPS)

14306 ceramic fibres (CF)

14305 cellular glass (CG)

14304 flexible elastomeric foam (FEF)

14303 mineral wool (MW)

Temperature (°C)

LNE   ASTM Workshop - NIST March 2012   5

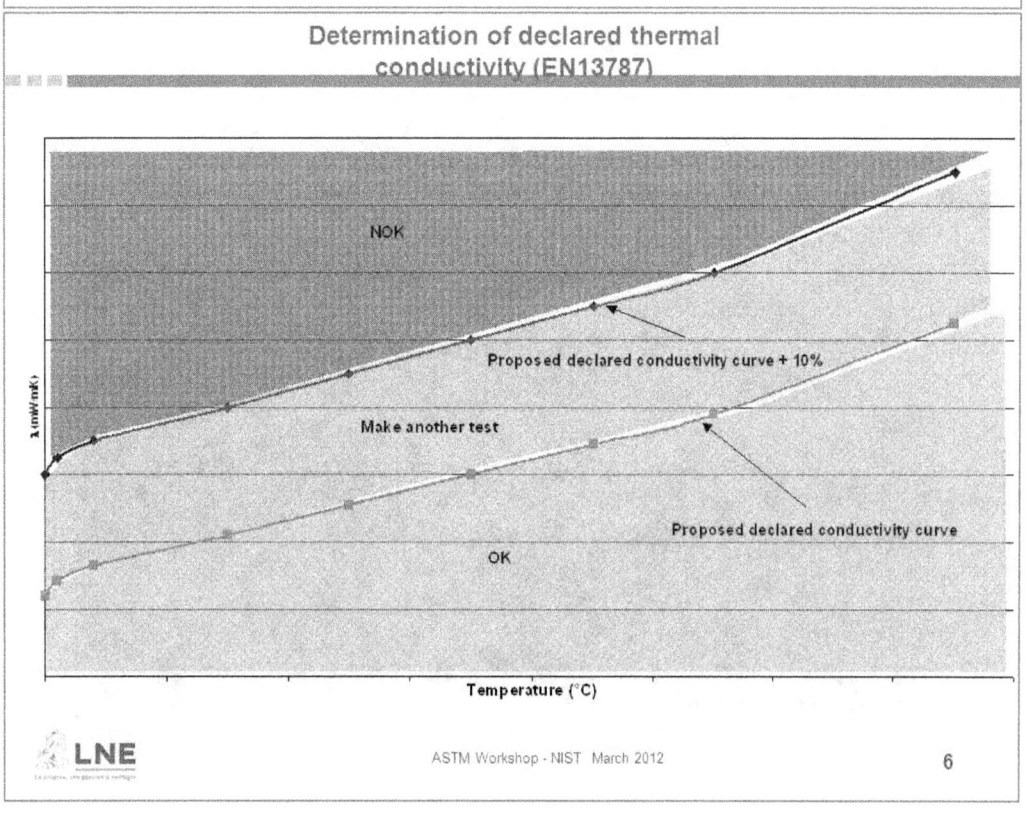

Determination of declared thermal conductivity (EN13787)

NOK

Proposed declared conductivity curve + 10%

Make another test

Proposed declared conductivity curve

OK

λ (mW/mK)

Temperature (°C)

LNE   ASTM Workshop - NIST March 2012   6

Intercomparison summary

Intercomparison at « room » temperature 1/3

- International Comparison on Thermal Conductivity Measurements of Insulating Materials by Guarded Hot Plate - CCT/P01 (organized by BIPM - CCT WG9)
    - ✓ 7 laboratories
    - ✓ 3 temperatures: 10, 23, 40°C

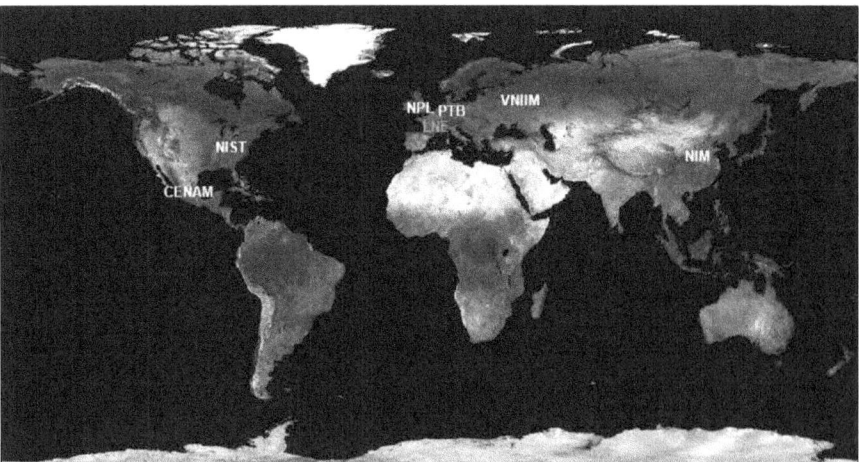

Analysis of results (1/2)

Repeated runs at 23 °C

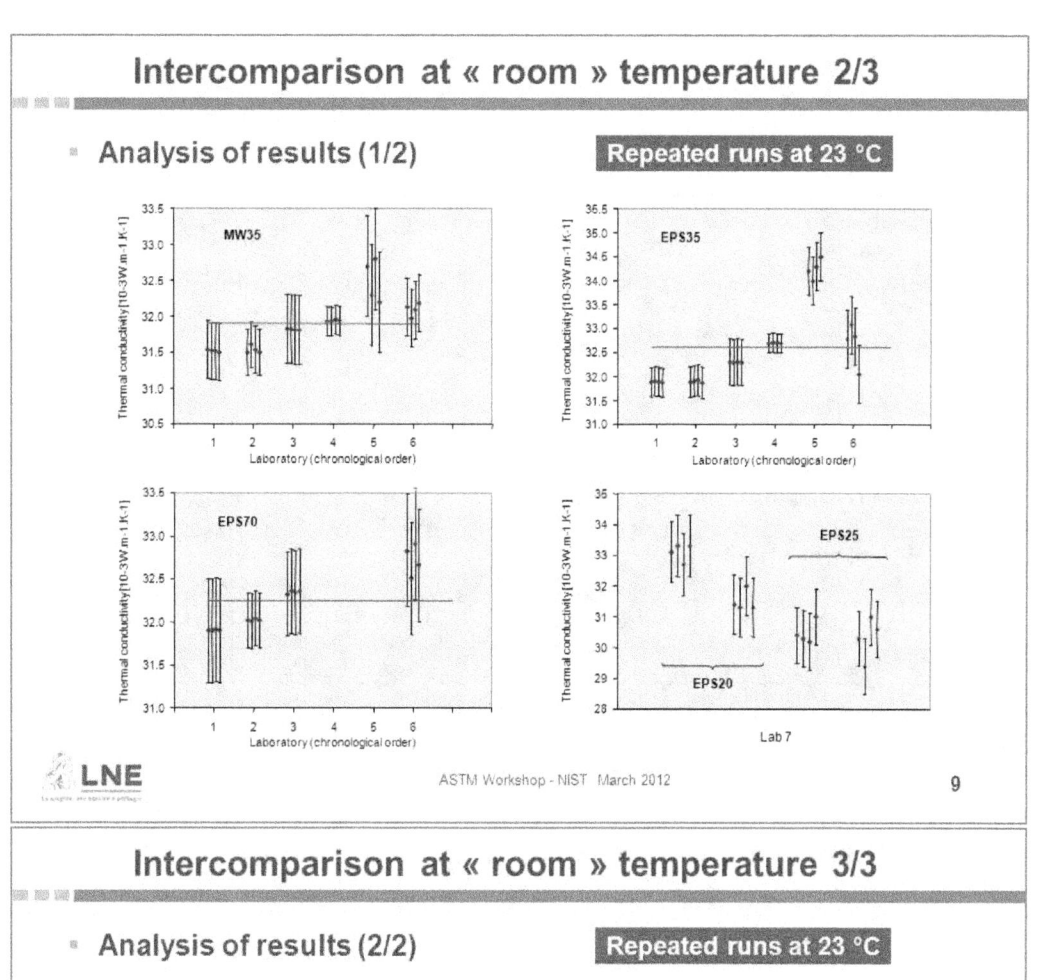

LNE

ASTM Workshop - NIST March 2012

9

Analysis of results (2/2)

Repeated runs at 23 °C

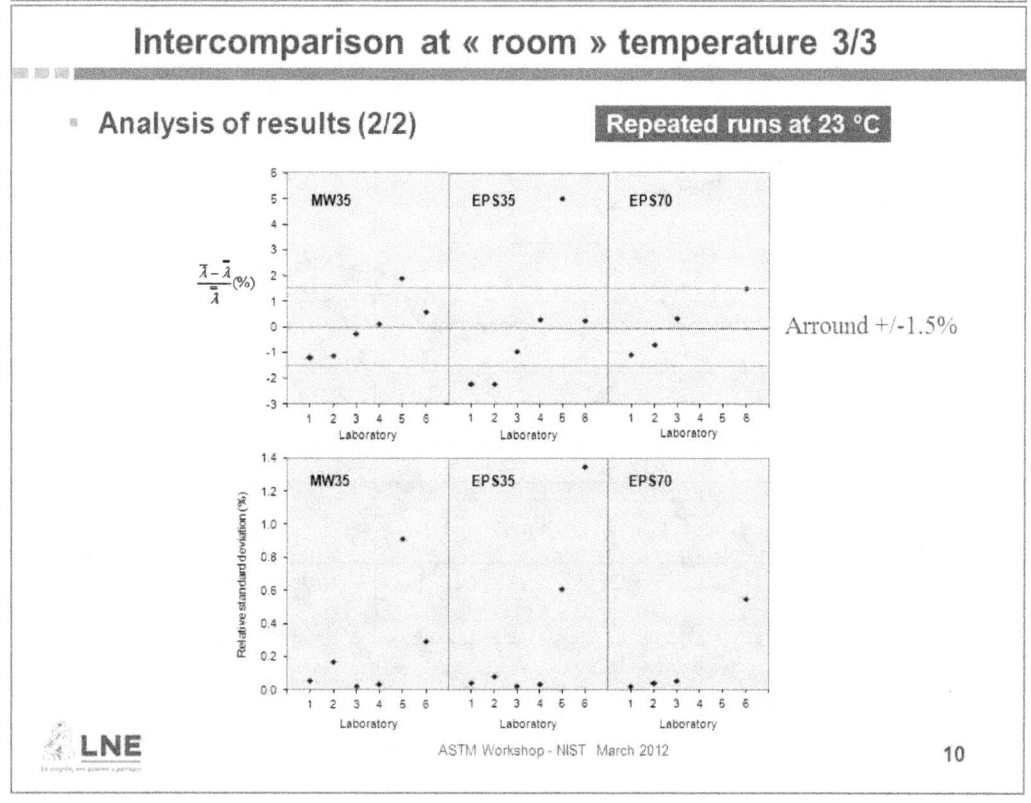

$$\frac{\overline{\lambda} - \overline{\overline{\lambda}}}{\overline{\overline{\lambda}}}(\%)$$

Arround +/-1.5%

LNE

ASTM Workshop - NIST March 2012

10

Intercomparison (cryogenic application) 1/3

Proceedings of the 30th International Thermal Conductivity Conference and the 18th International Thermal Expansion Symposium

NPL   Padova

TABLE I COMPARISON OF NPL AND DFT VALUES FOR IRMM-440

NPL LTGHP Mean Specimen Temperature °C	NPL LTGHP Measured Thermal Conductivity $W \cdot m^{-1} \cdot K^{-1}$	DFT Thermal Conductivity Reference $W \cdot m^{-1} \cdot K^{-1}$	Difference (Measured - Reference) (%)
-174.7	0.0117	0.0110	7.0
-149.2	0.0139	0.0137	1.6
-97.2	0.0190	0.0192	-1.4
-49.1	0.0245	0.0244	0.3
0.2	0.0296	0.0296	-0.1
50.4	0.0355	0.0350	1.4

These values are not part of the official material certification

Intercomparison (cryogenic application) 2/3

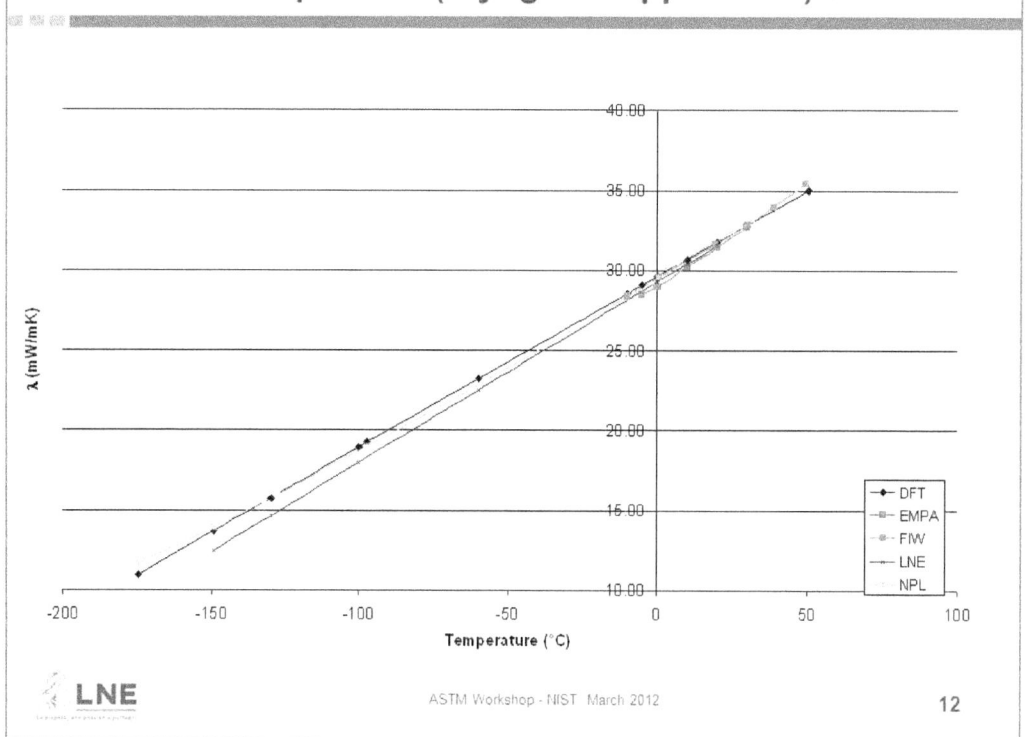

Intercomparison (cryogenic application) 3/3

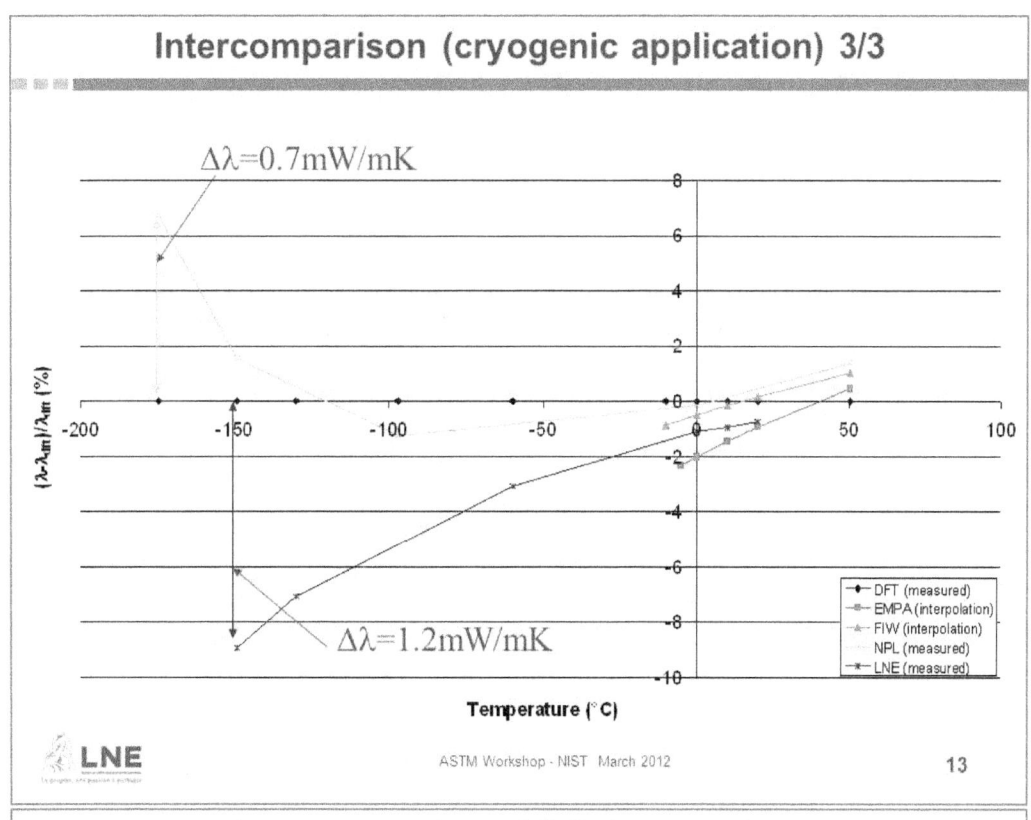

Intercomparison high temperature

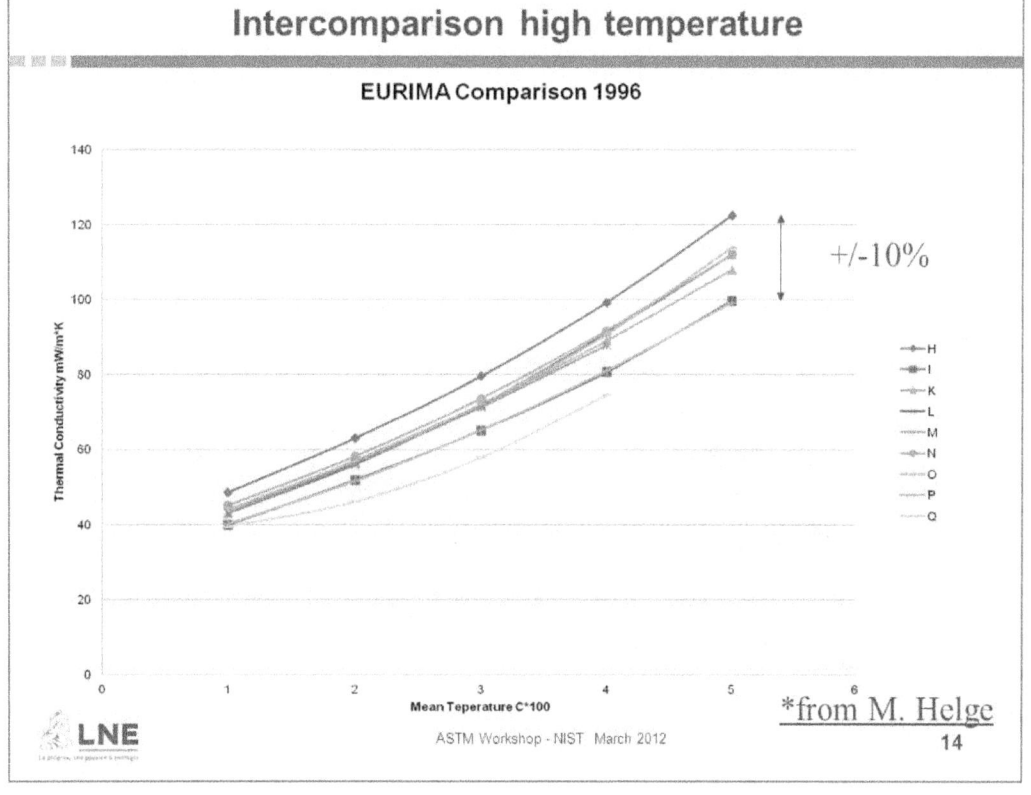

*from M. Helge

REFERENCE MATERIAL

Current Situation

 **Available reference materials**

- ✓ Low temperature (below 0°C till -170°C)
    - None
- ✓ Building temperature (0°C, 60°C)
    - IRMM440 ➜ MW
    - NIST ➜ MW
- ✓ High temperature (above 60°C till +700°C or more)
    - None

Difficulties

- Low Temperature
  - ✓ At low temperature thermal conductivity of material can be very low
    - At 10°C GHP is better than +/-1.5% ➔ +/- 0.4mW/mK on MW
    - With the same accuracy and same material At –170°C +/-1.5% means 0.15mW/mK.
  - ✓ Gas phase change
- High temperature
  - ✓ Radiative component of thermal transmission not negligible
- In all cases
  - ✓ Temperature sensors should be calibrated on a large range of temperature
    - Technical difficulties
    - Polynomial definition

LNE EQUIPMENT
For
Low and High temperature

NETZSCH GHP (low and high temperatures)

2 samples: 500x500mm
Tmin: -170°C
Tmax: 700°C
Thickness: 10 – 70 mm

Cooling fluide
$\begin{cases} \text{LN2 : -170, 0°C} \\ \text{Ethanol: 0, 20°C} \\ \text{Air : 20°C, 700°C} \end{cases}$

LN2 Tanker

LNE

High temperature GHP

2 samples: 300x300mm
Tmin: 50°C
Tmax: 400°C
Thickness : 20 – 50mm

Homemade GHP

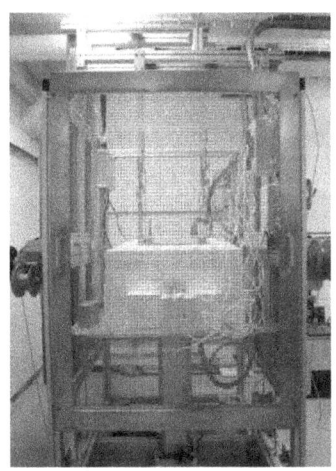

LNE

Conclusion

- There is important needs for insulation reference materials
  - ✓ For low temperature
    - Cryogenic application between −170°C and 0°C
  - ✓ For high temperature
    - 60°C – 700°C

- Need to find materials with proper features to be used as standards

- Need round robin tests to validate measurements

NIST 500 mm Guarded-Hot-Plate Apparatus

**2nd Operators Workshop on High-Temperature
Guarded-Hot-Plate and Pipe Measurements**

March 19-20, 2012

Construction Photograph Record

Robert Zarr and John Hettenhouser

National Institute of Standards and Technology
100 Bureau Drive
Gaithersburg, Maryland, United States

Superstructure and
Vacuum Cart

NIST Welding Shop

Installation of
Vacuum Baseplate and Bell Jar

Building 226, Room B107

Installation of
Vacuum Baseplate and Bell Jar

Building 226, Room B107

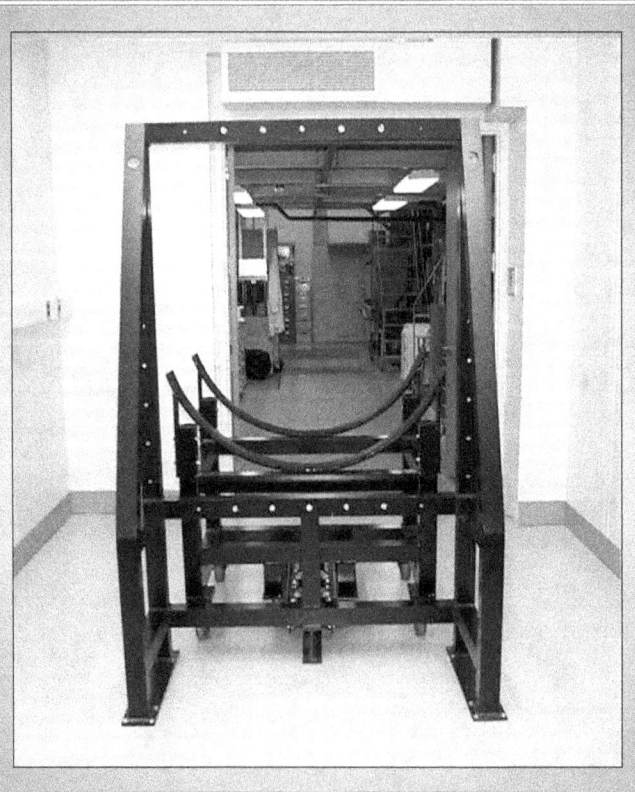

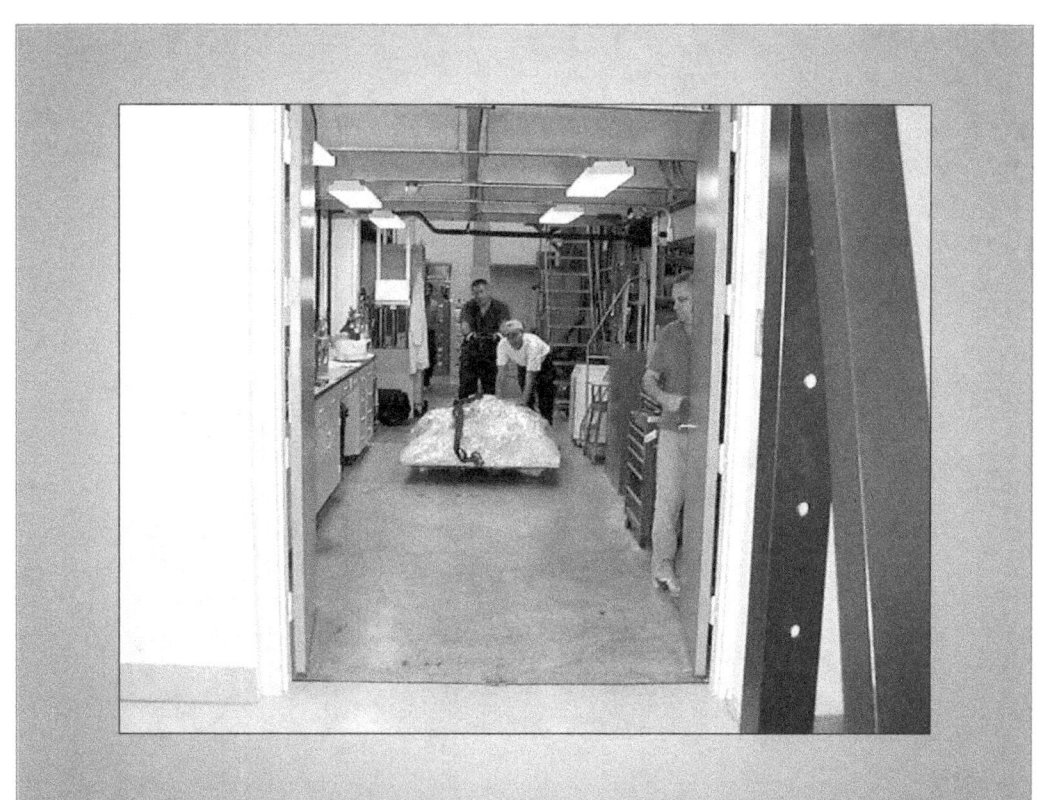

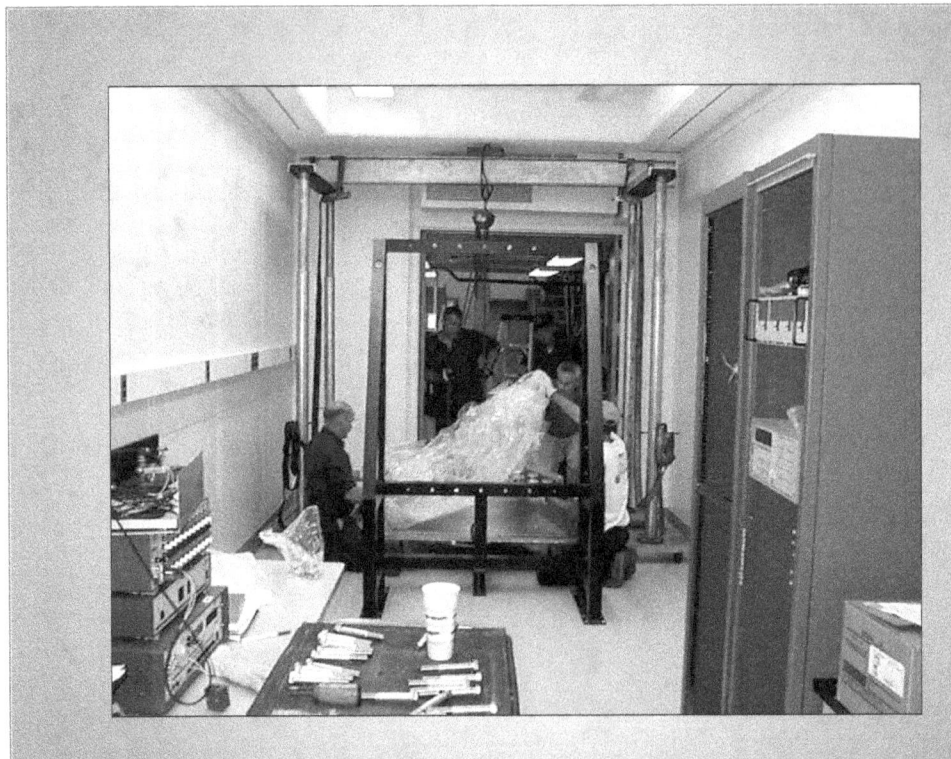

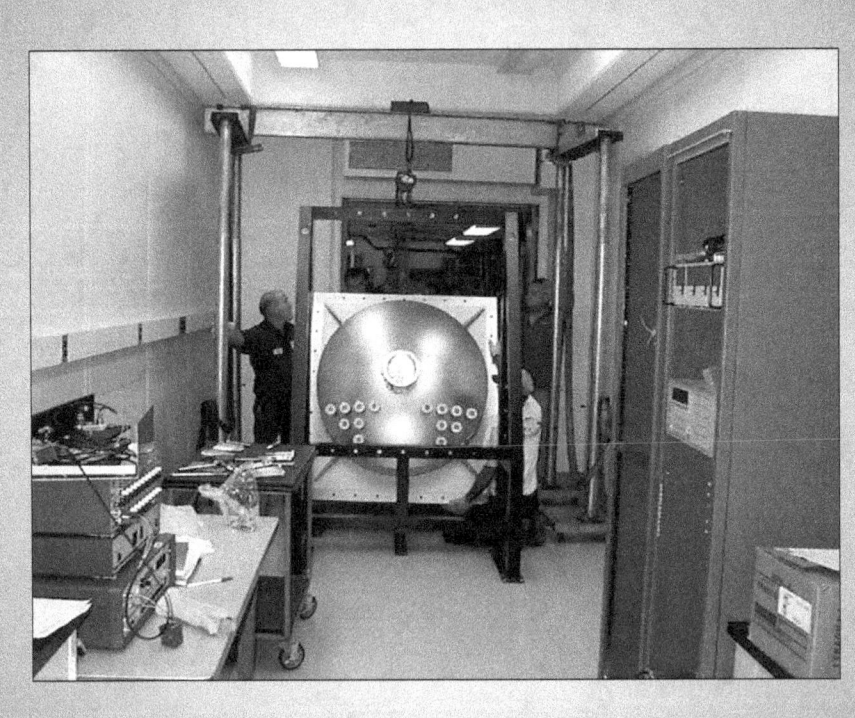

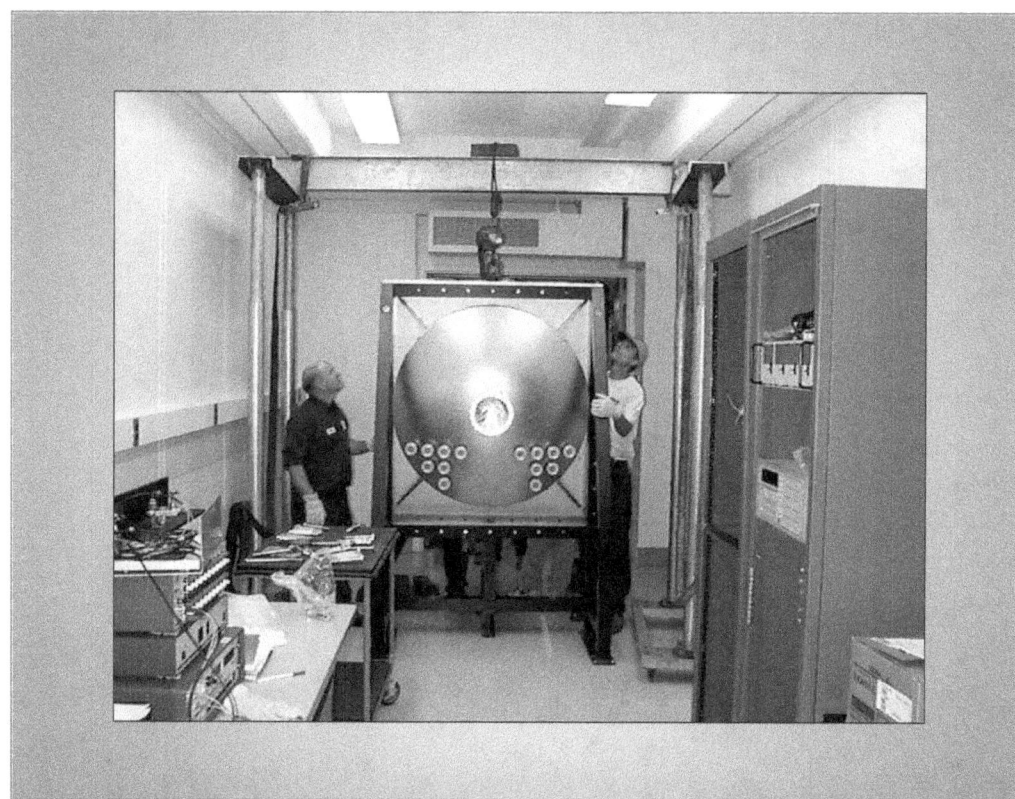

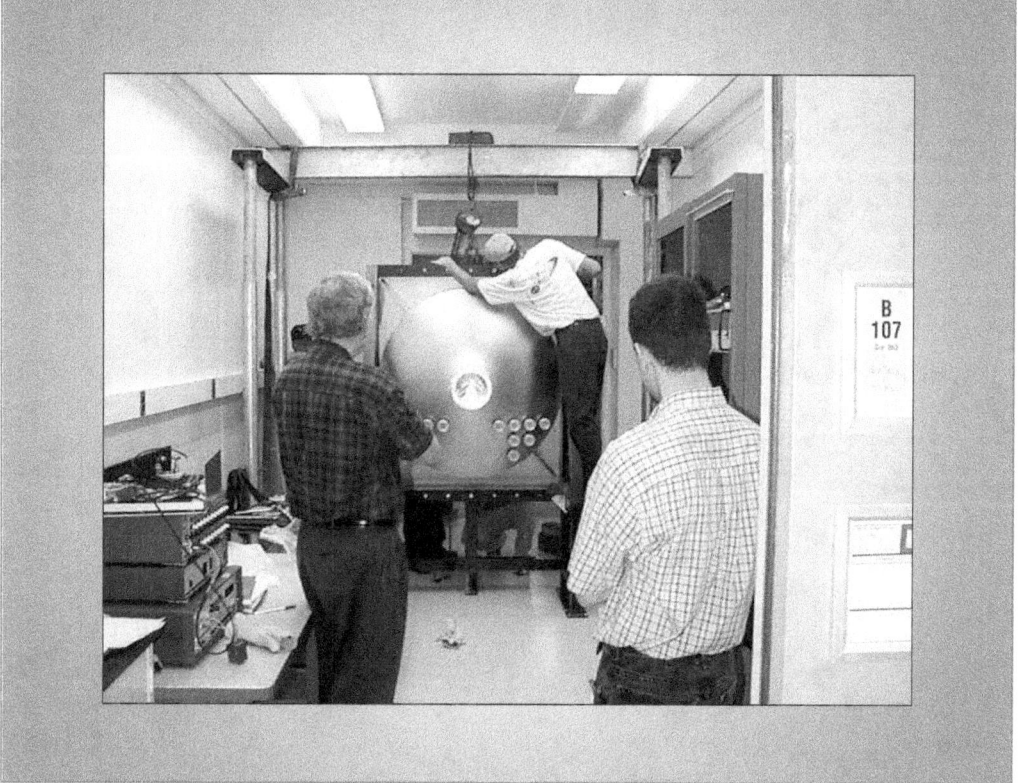

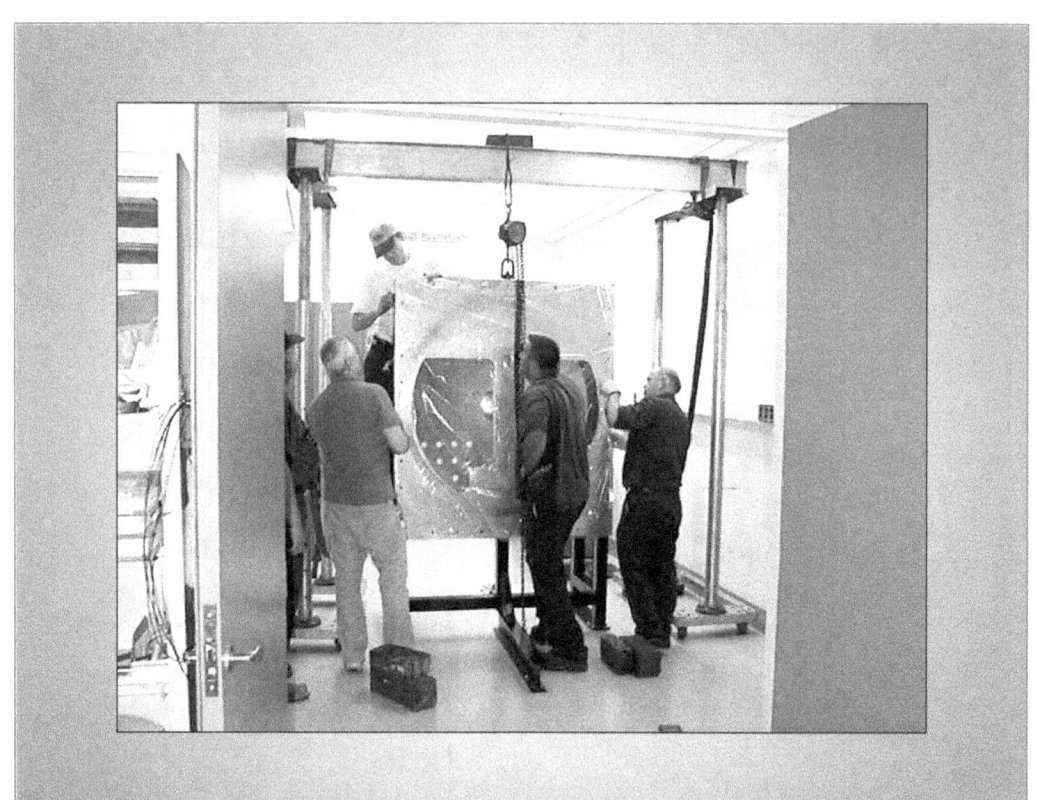

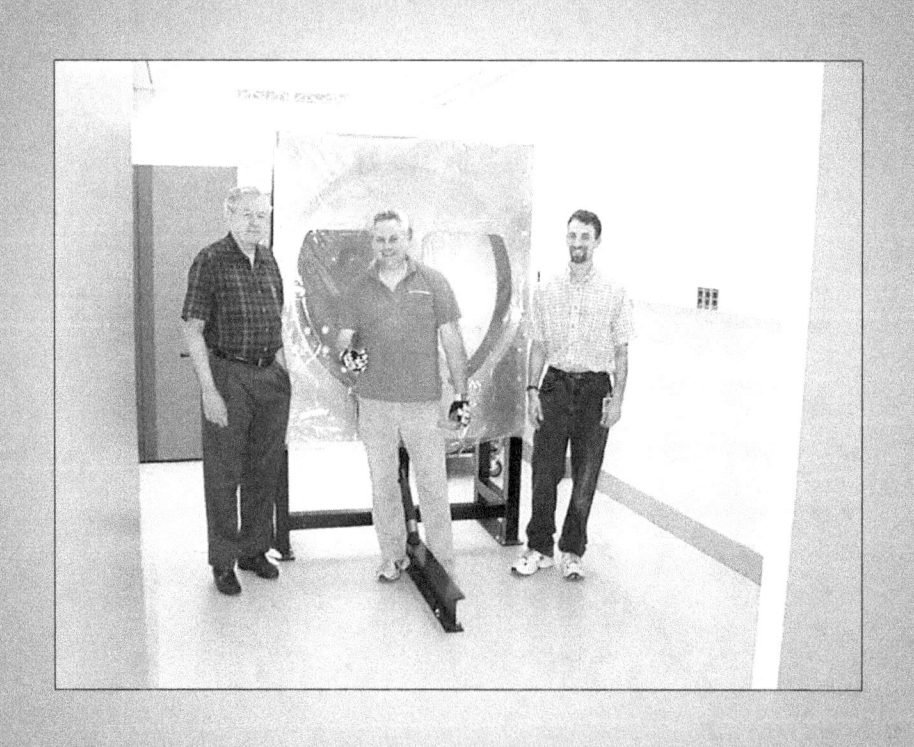

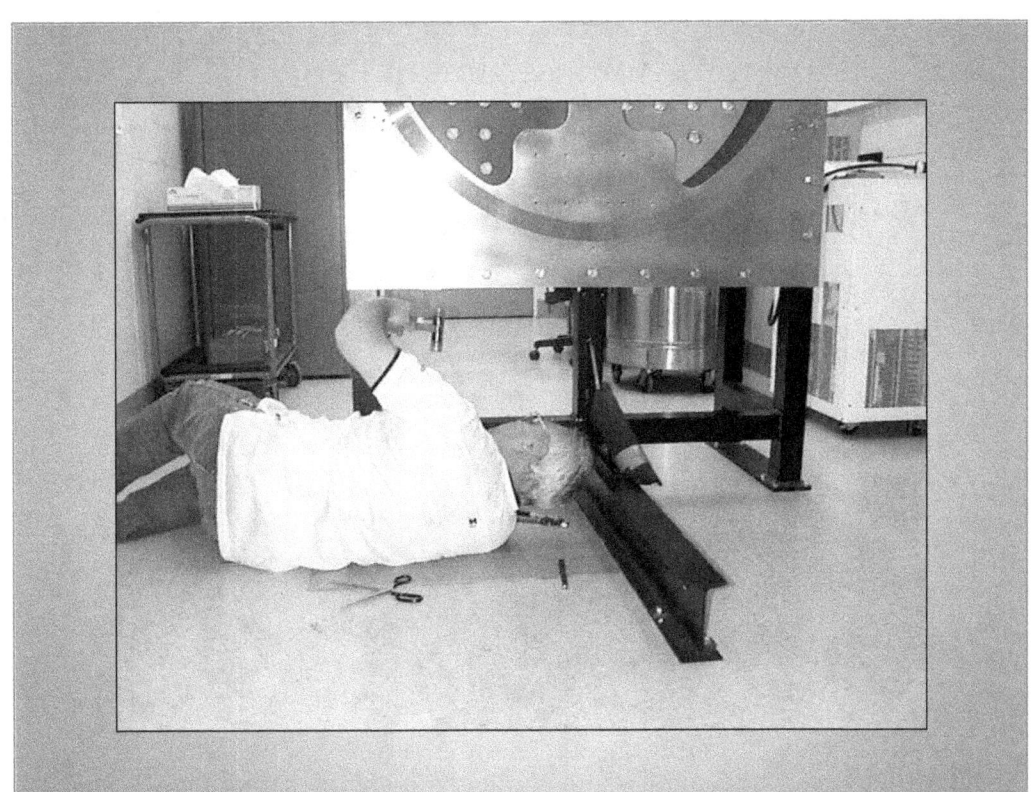

Vacuum System

Building 226, Room B107

Frame for Instrument Panel

Instrument Panel

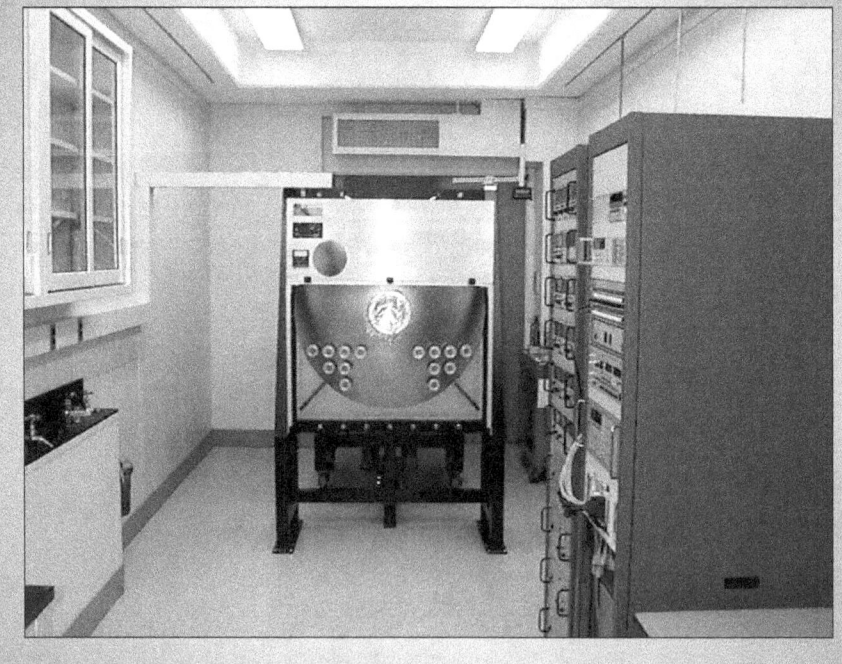

Diffusion Pump

Helium Leak Check

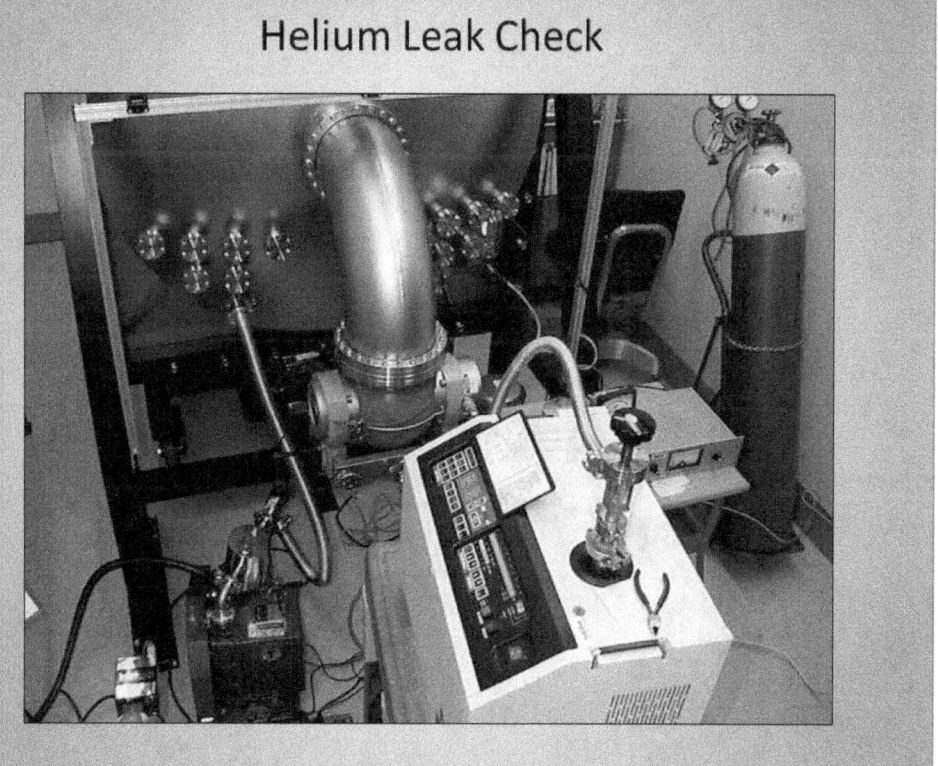

Helium Leak Check

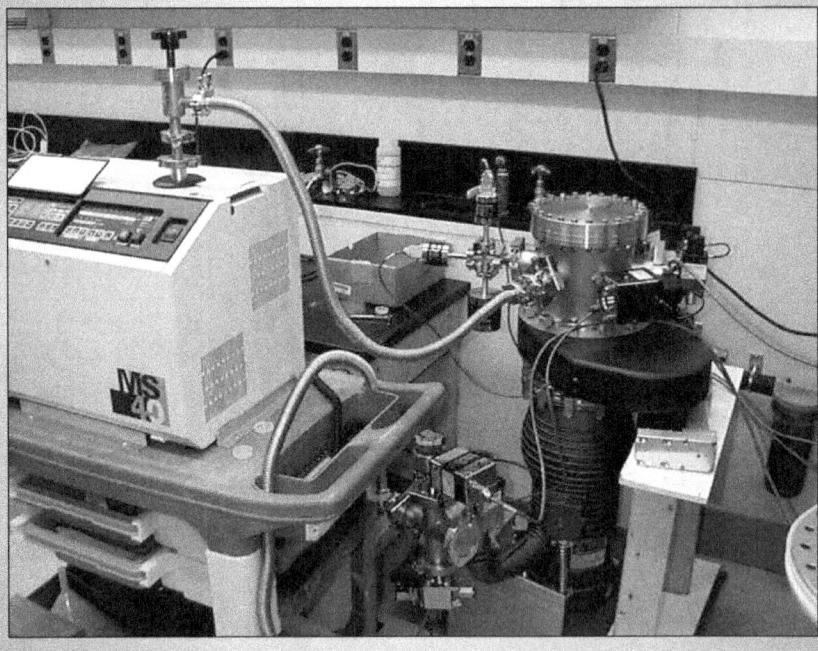

Source of Leak

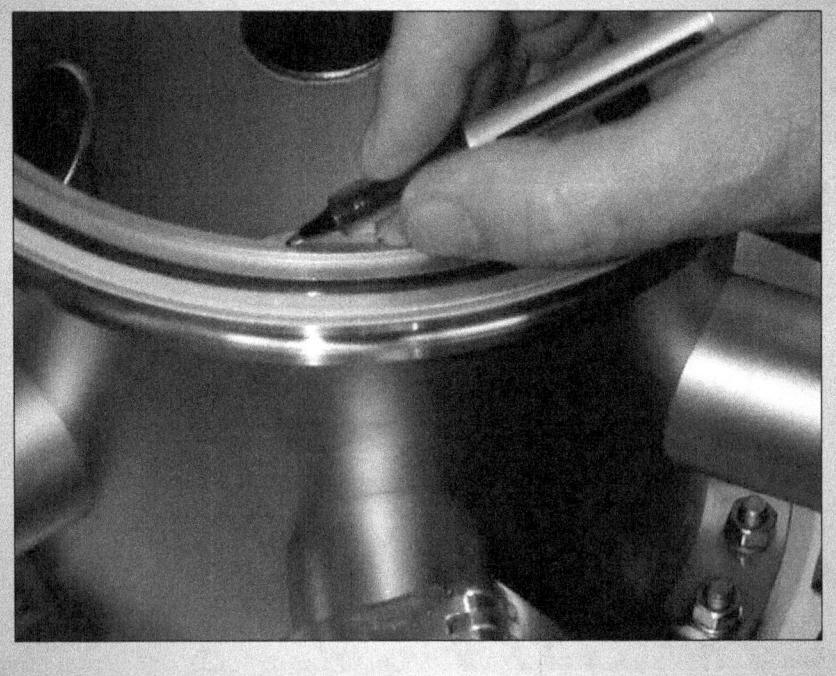

Installation of Overhead Rails

Building 226, Room B107

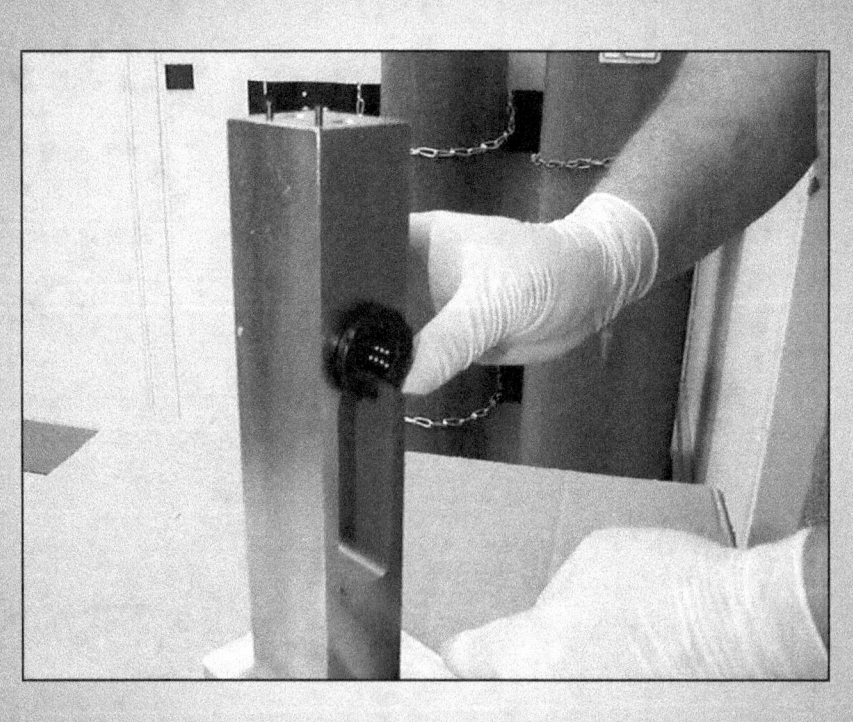

Hot Plate Components Fabrication

NIST Machine Shop
and
NIST EDM Shop

Meter-plate Heater

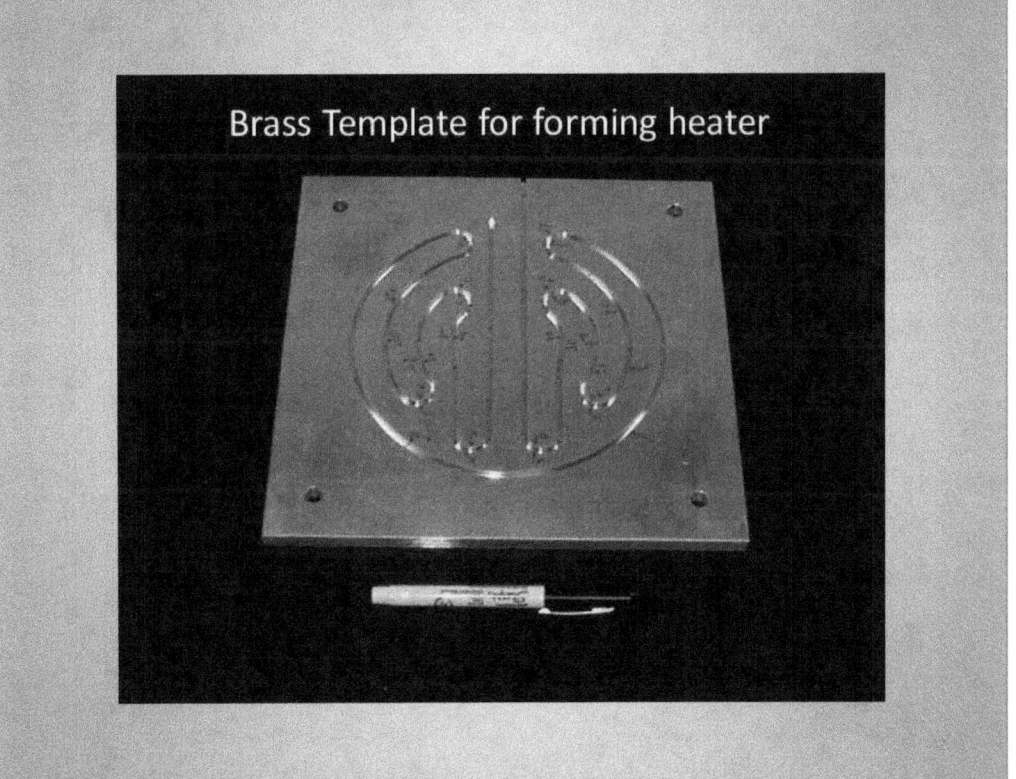

Brass Template for forming heater

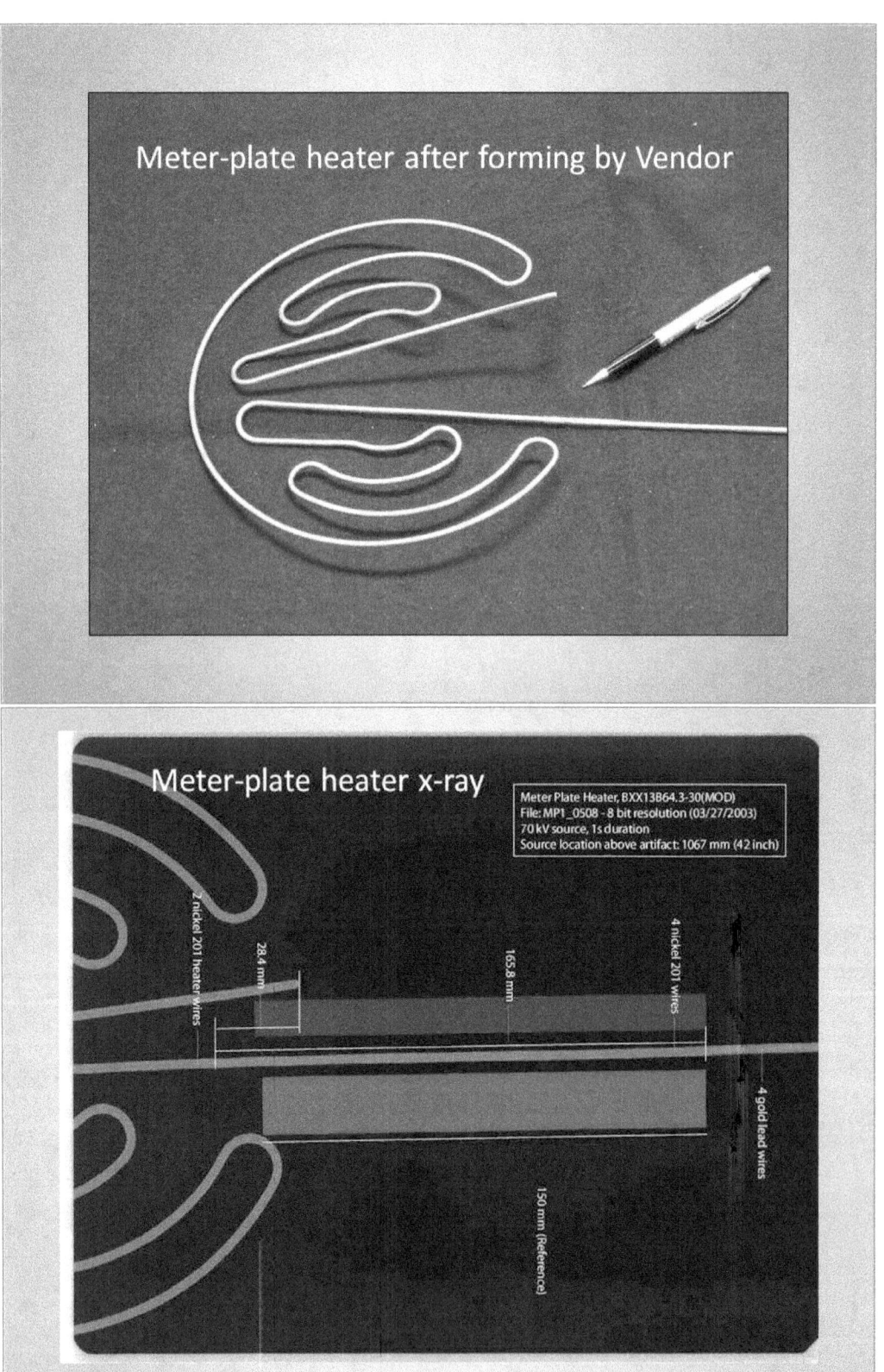

Hot Plate Construction

Vacuum Brazing

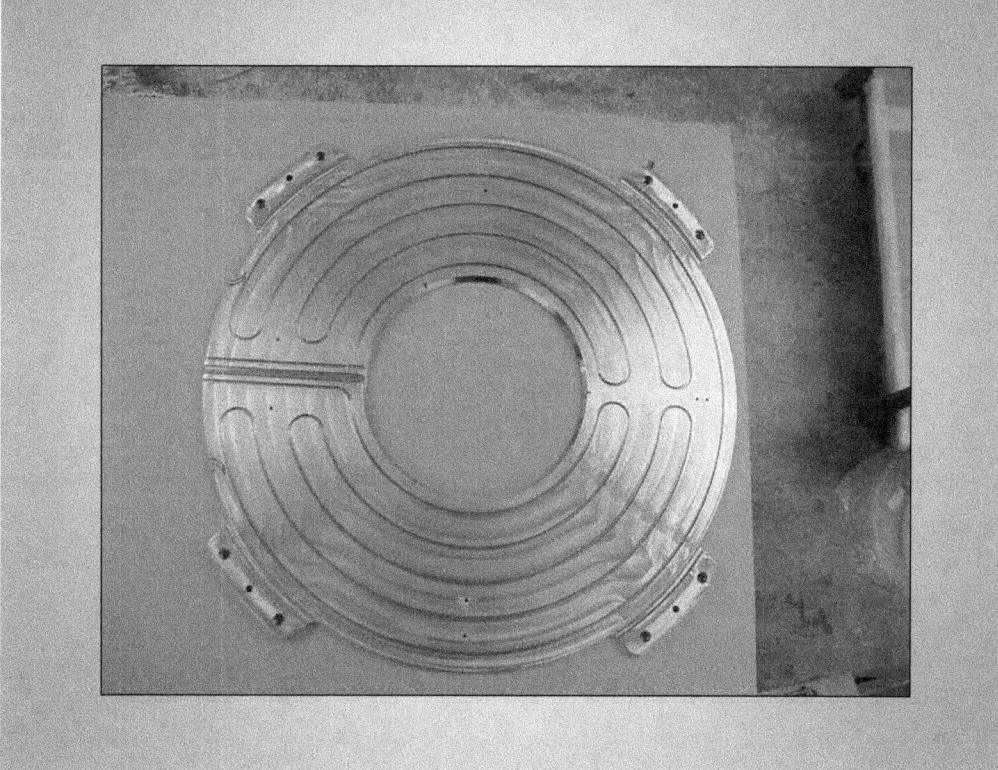

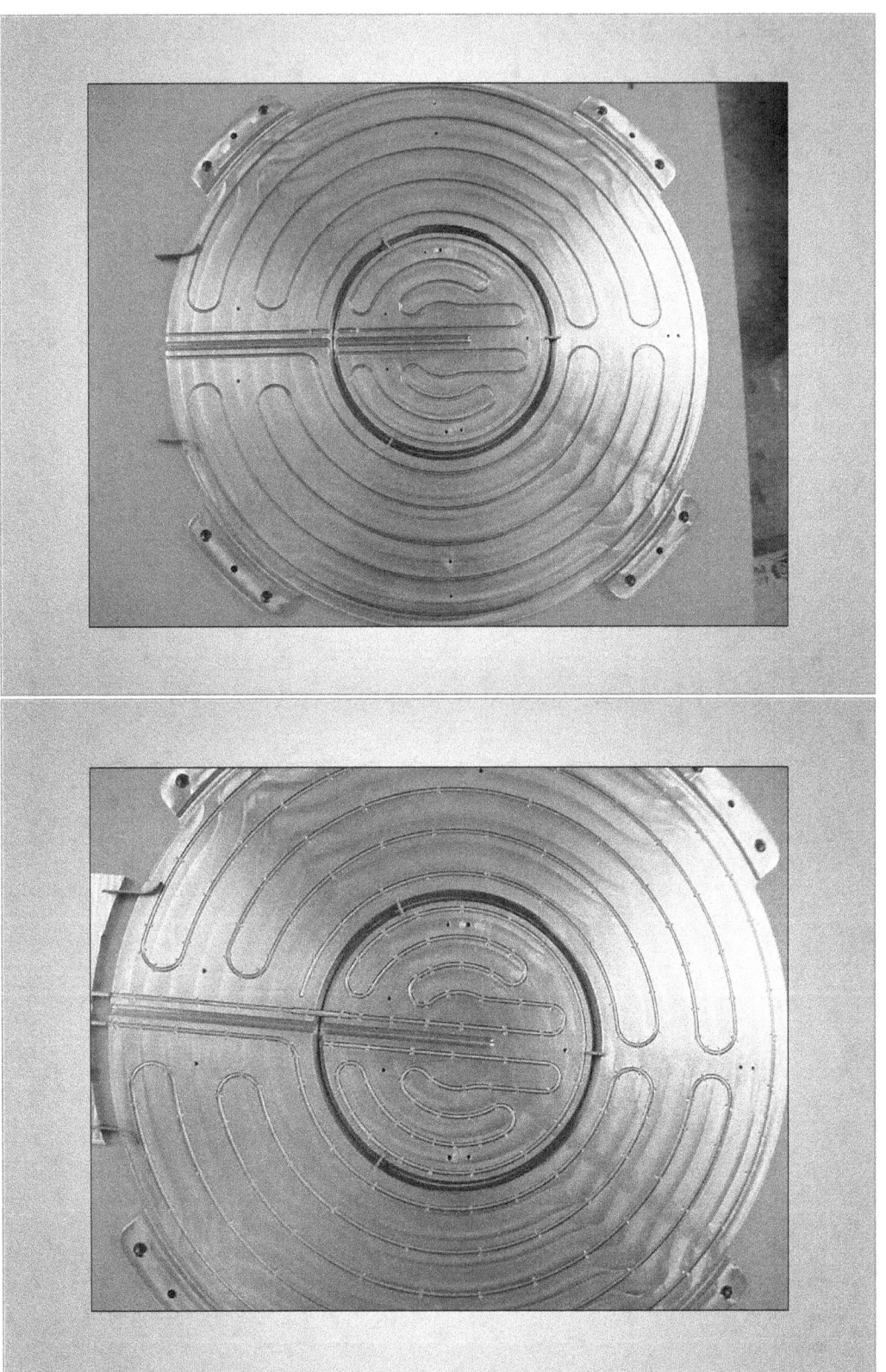

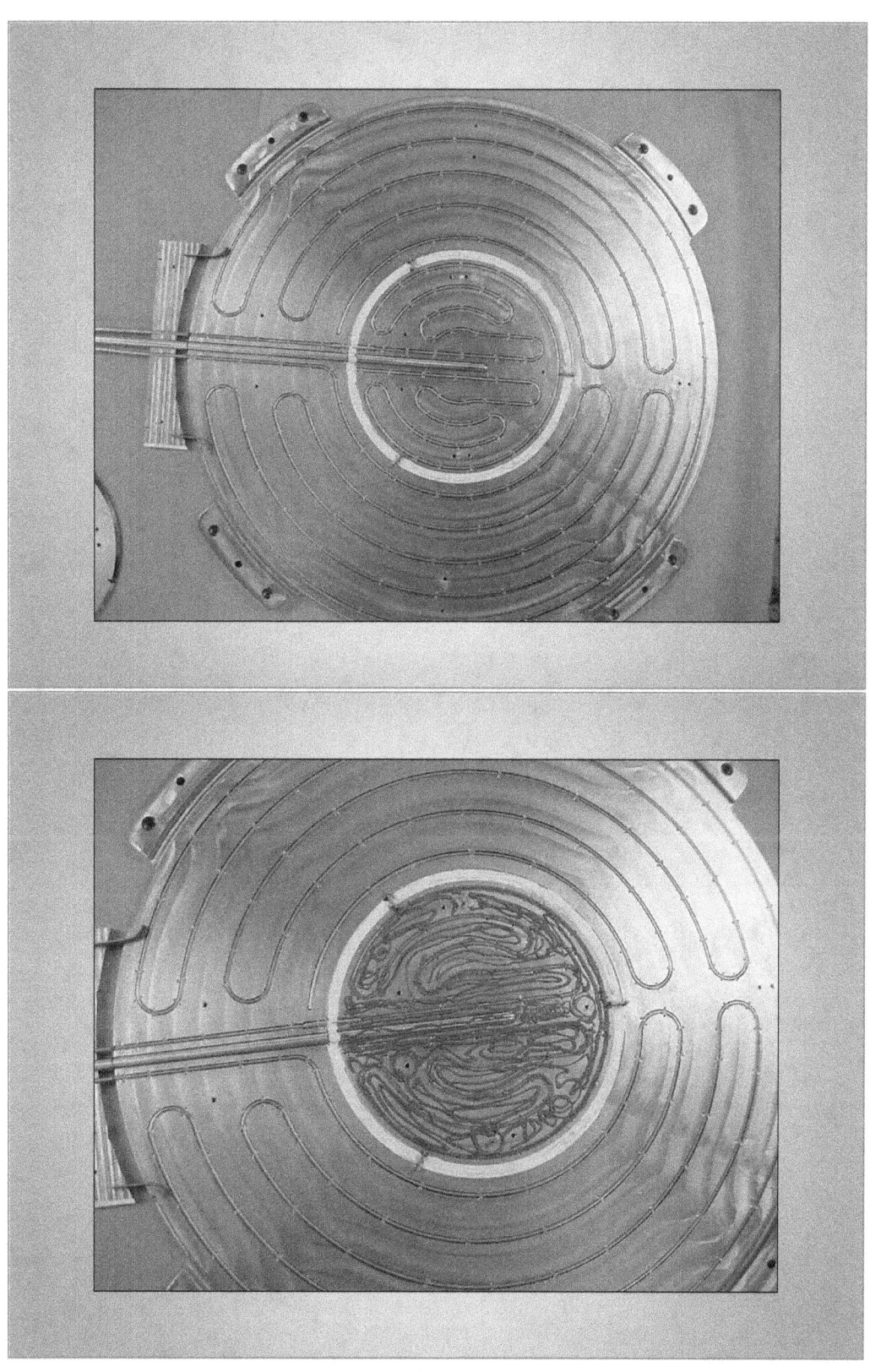

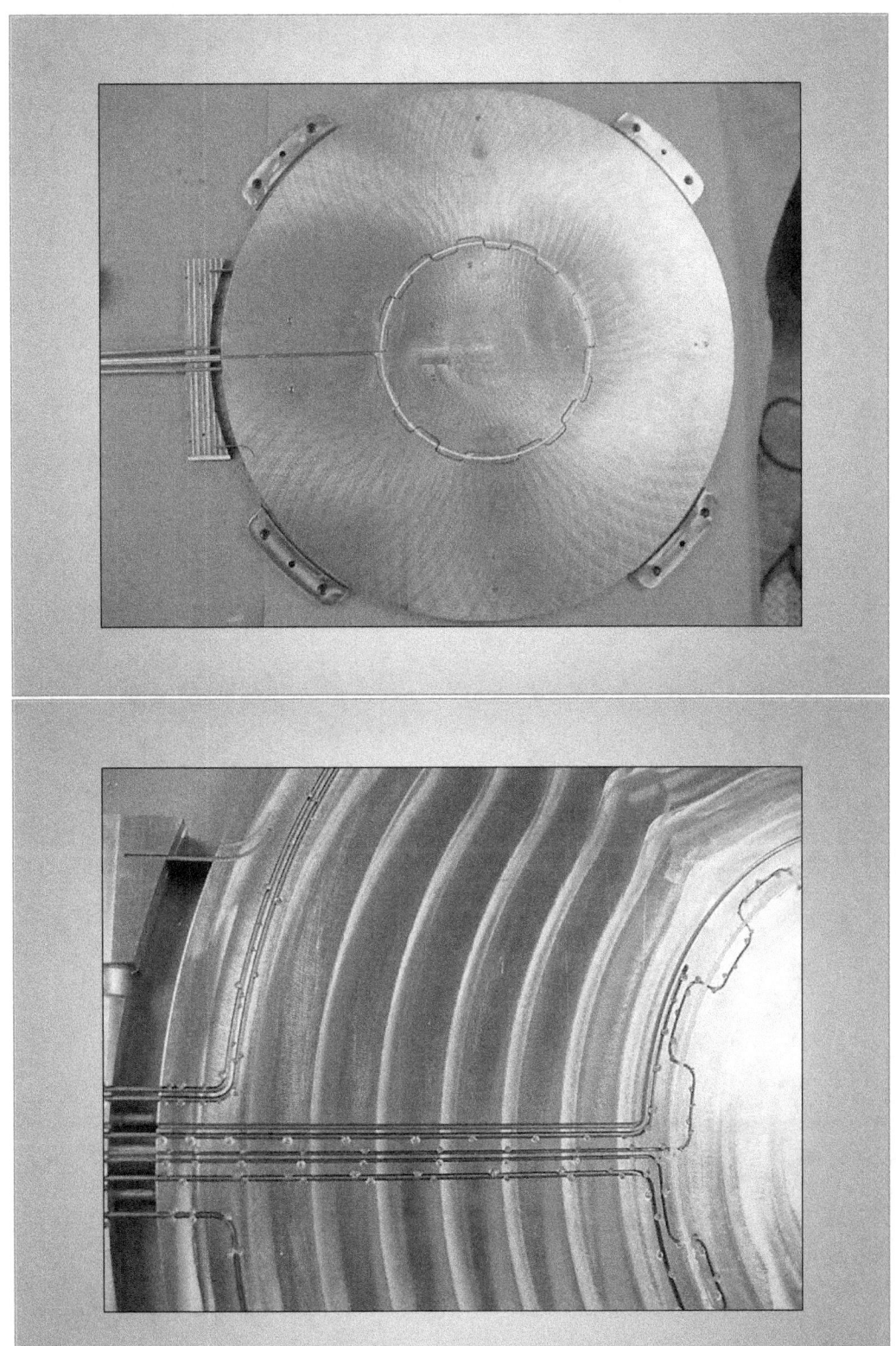

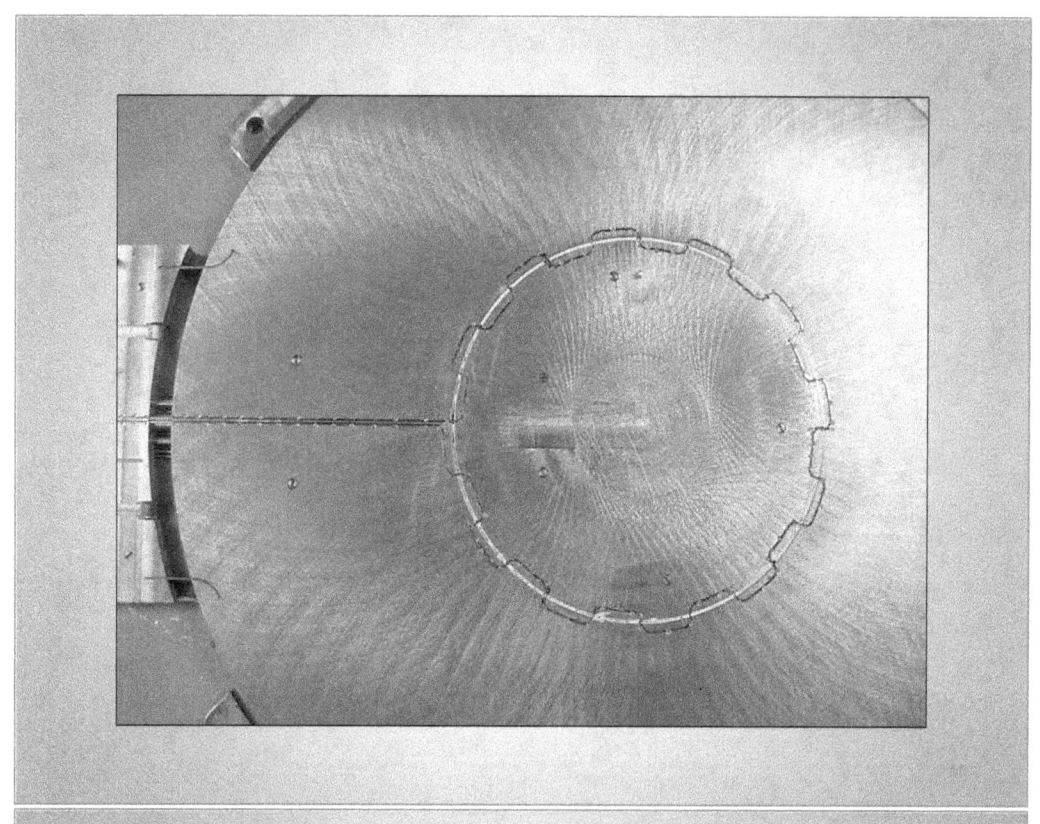

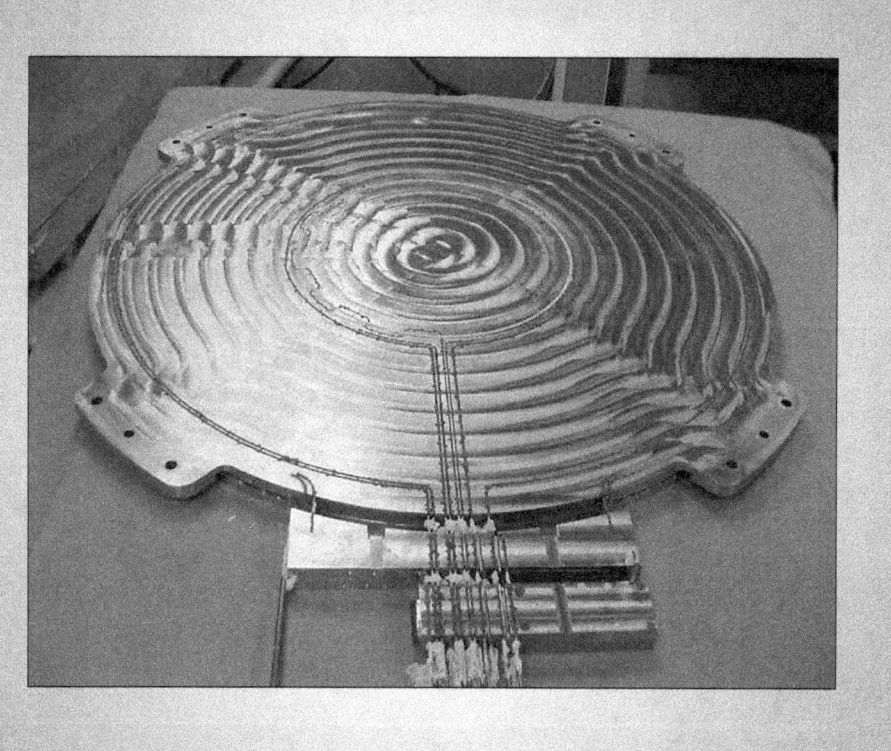

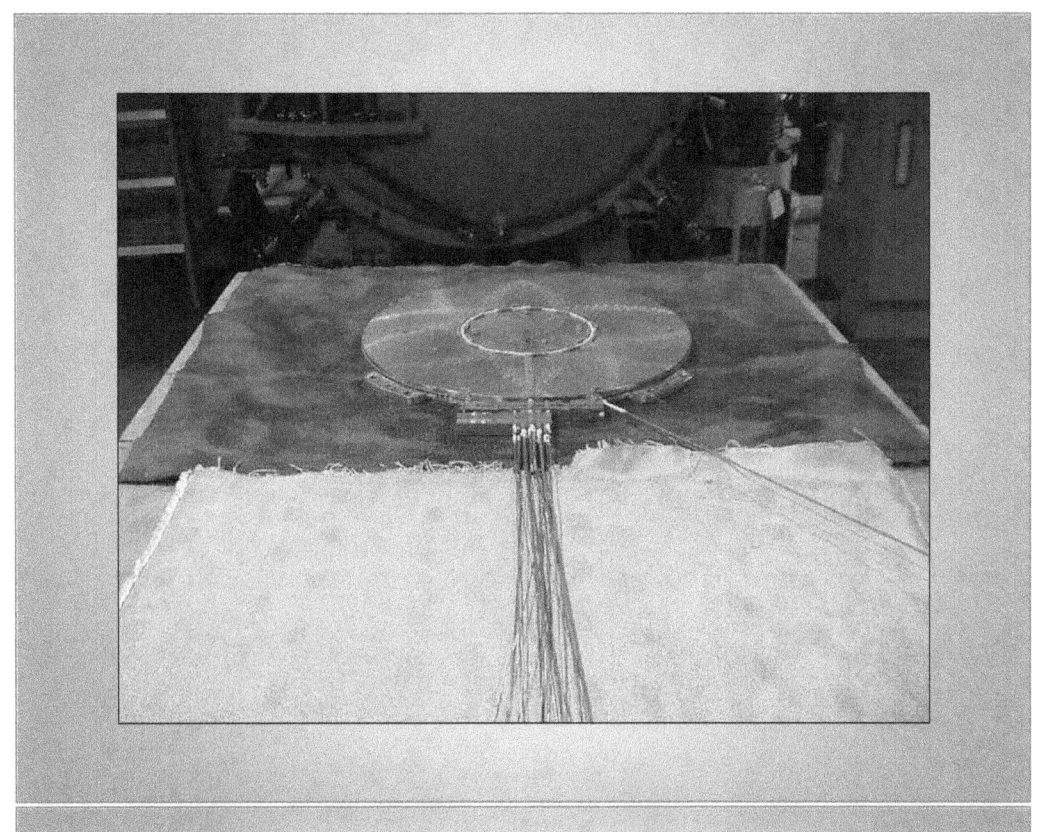

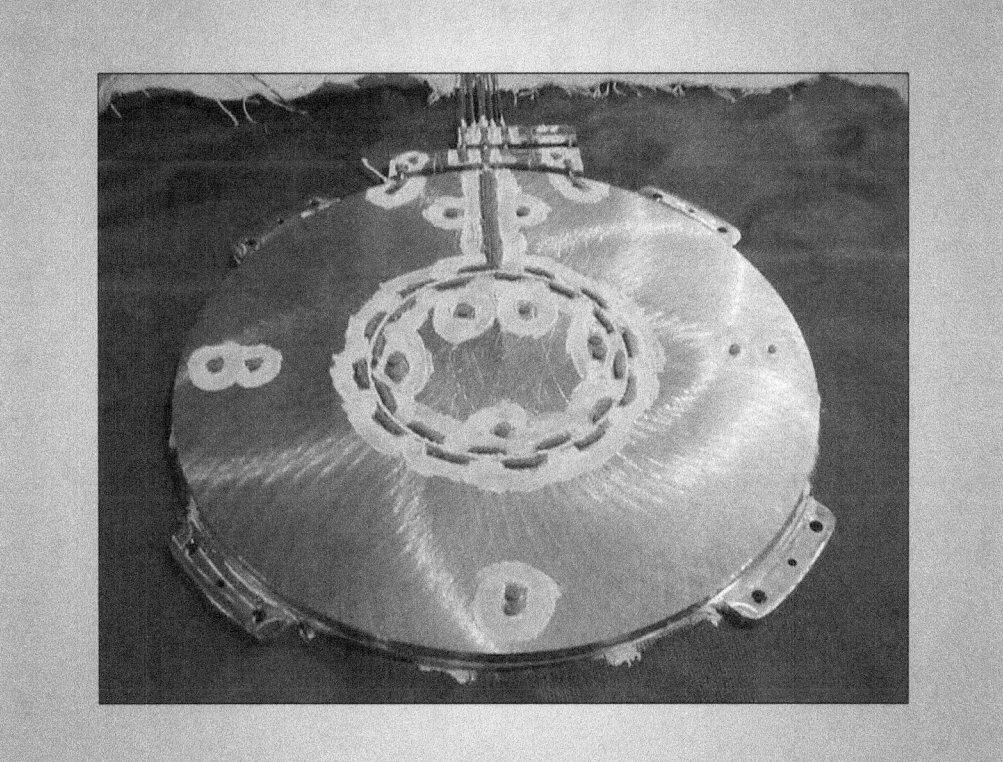

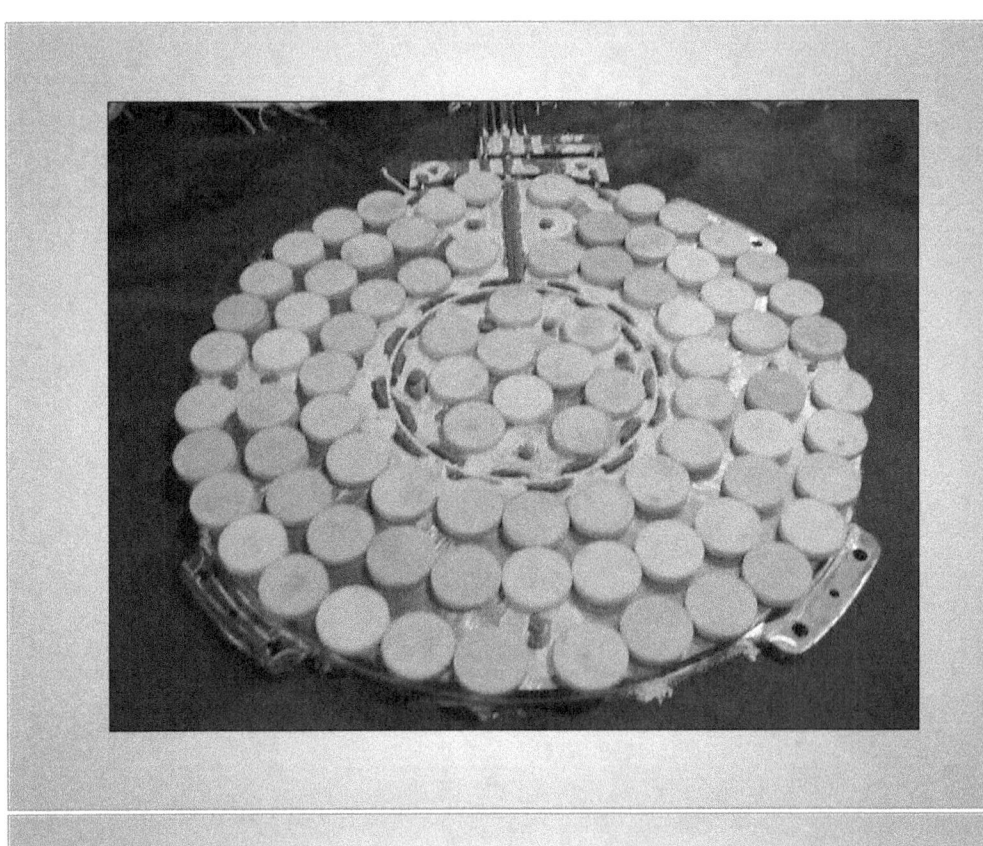

Hot Plate Construction

Post brazing
NIST Machine Shop

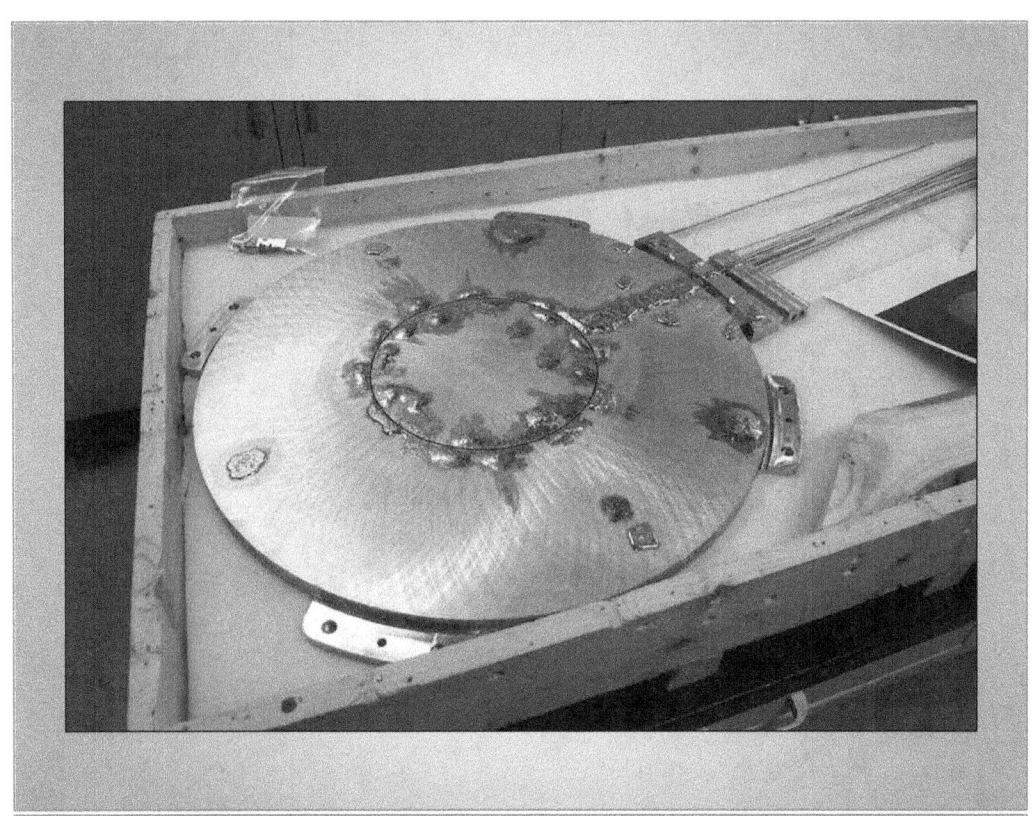

103

Dimensional Inspection

NIST Machine Shop
Quality Inspection CMM
(Coordinate Measurement Machine)

Dimensional Reference
(Baseline)

NIST Dimensional Metrology
Moore M48 CMM

Cold Plate Assemblies – Thermometer Plate, Heater Plate, and Coolant Plate

Original Sketches, Fabrication, Vacuum Brazing, Post-brazing Machining, Coating

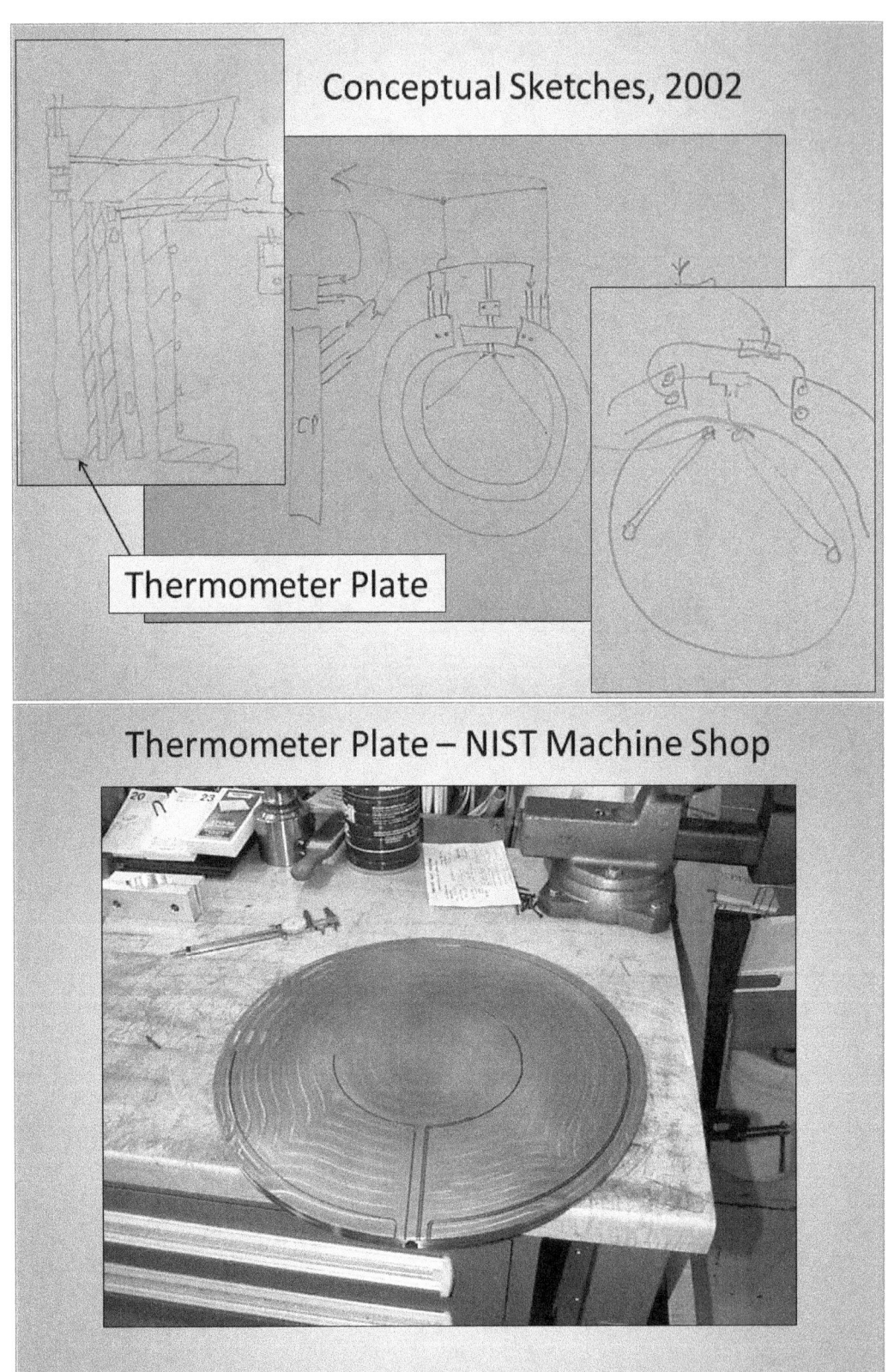

Conceptual Sketches, 2002

Thermometer Plate

Thermometer Plate – NIST Machine Shop

Thermometer Plate – Vacuum Brazing, Side 1

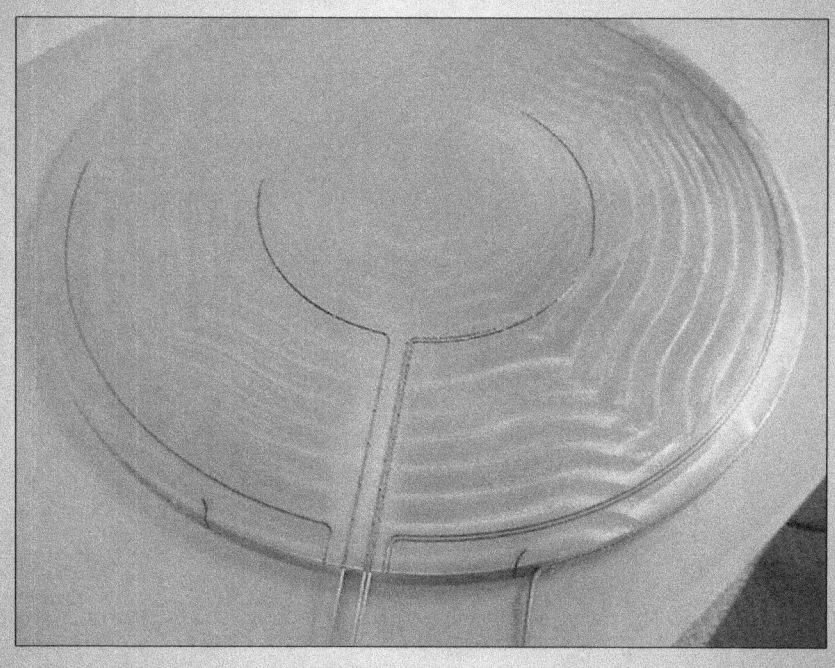

Thermometer Plate – Vacuum Brazing, Side 1

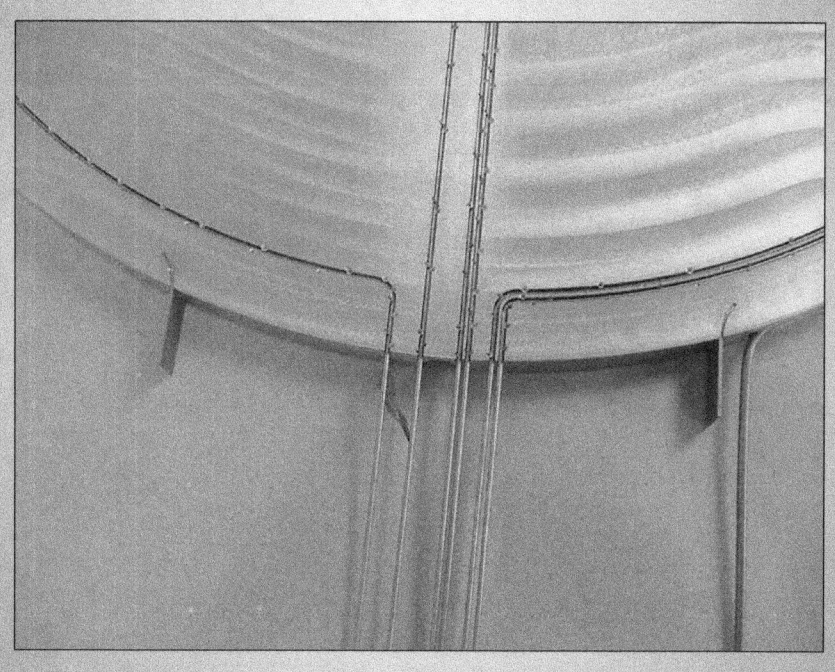

Thermometer Plate – Vacuum Brazing, Side 1

Thermometer Plate – Vacuum Brazing, Side 1

Thermometer Plate – Vacuum Brazing, Side 1

Thermometer Plate – Vacuum Brazing, Side 2

Thermometer Plate – Return to NIST

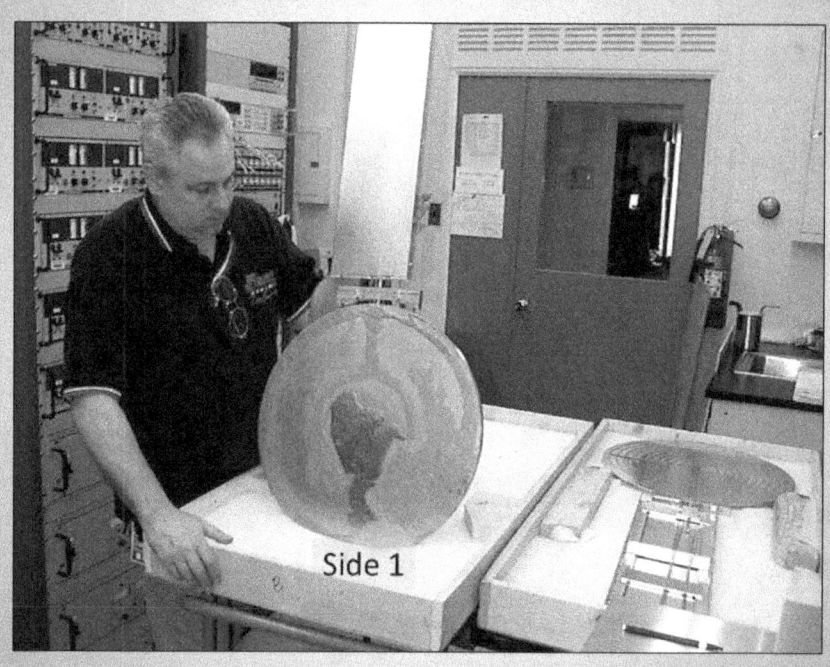

Side 1

Thermometer Plate – Return to NIST

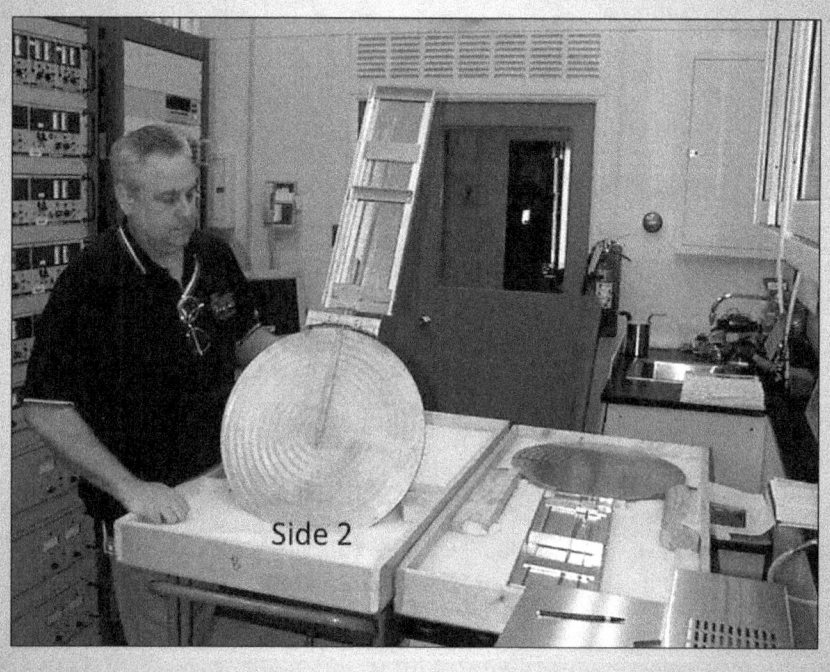

Side 2

Thermometer Plate – Excess Braze Removal

Thermometer Plate – Coating Vendor

Thermometer Plate – Coating Vendor

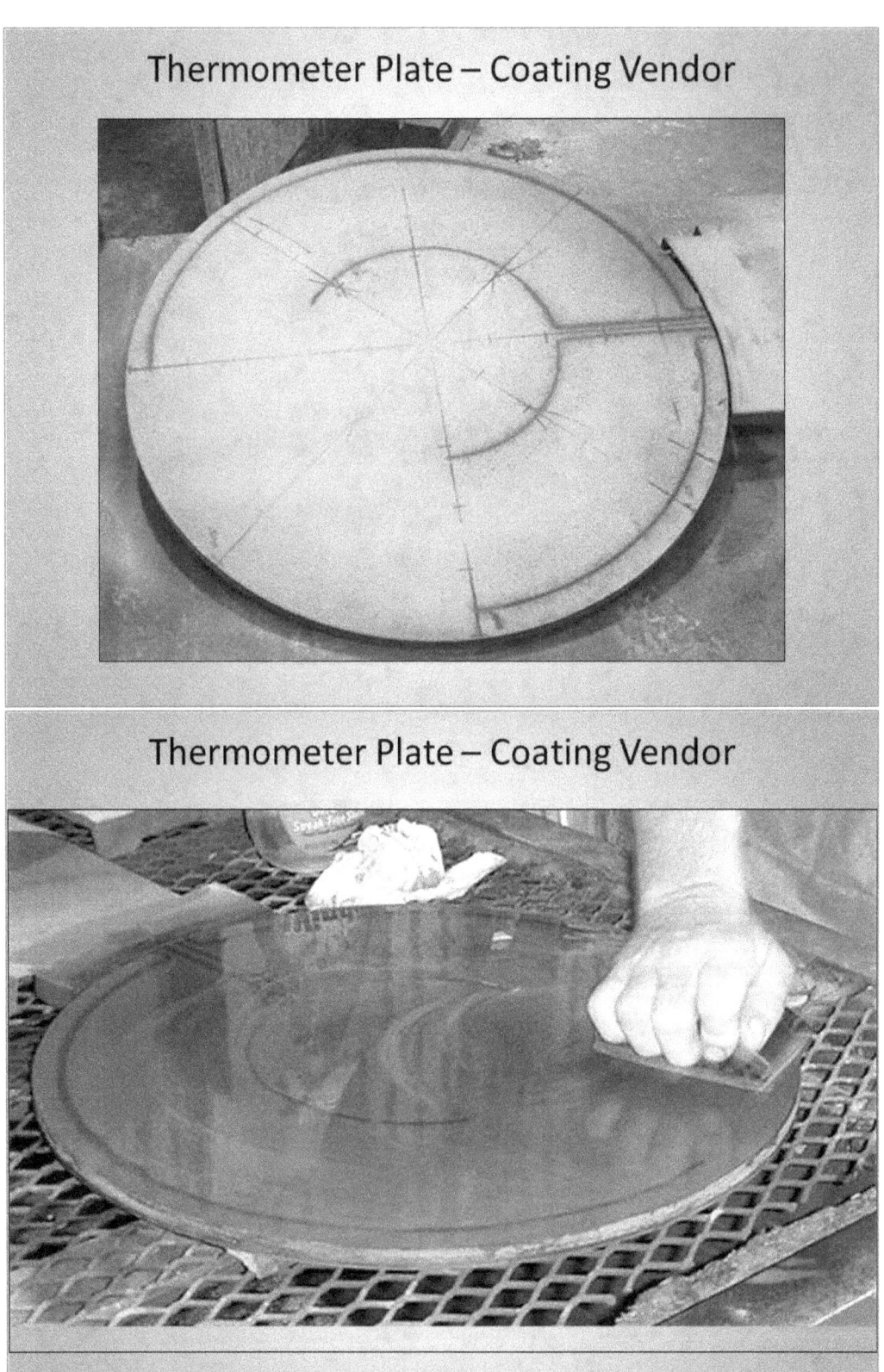

Thermometer Plate – Coating Vendor

Thermometer Plate – Coating Vendor

Cold Plate Heater

Cold Plate Heater – Vacuum Brazing

Cold Plate Heater – Vacuum Brazing

Cold Plate Heater – Vacuum Brazing

Coolant Plate (Spiral Channel HX)

Coolant Plate (Spiral Channel HX)

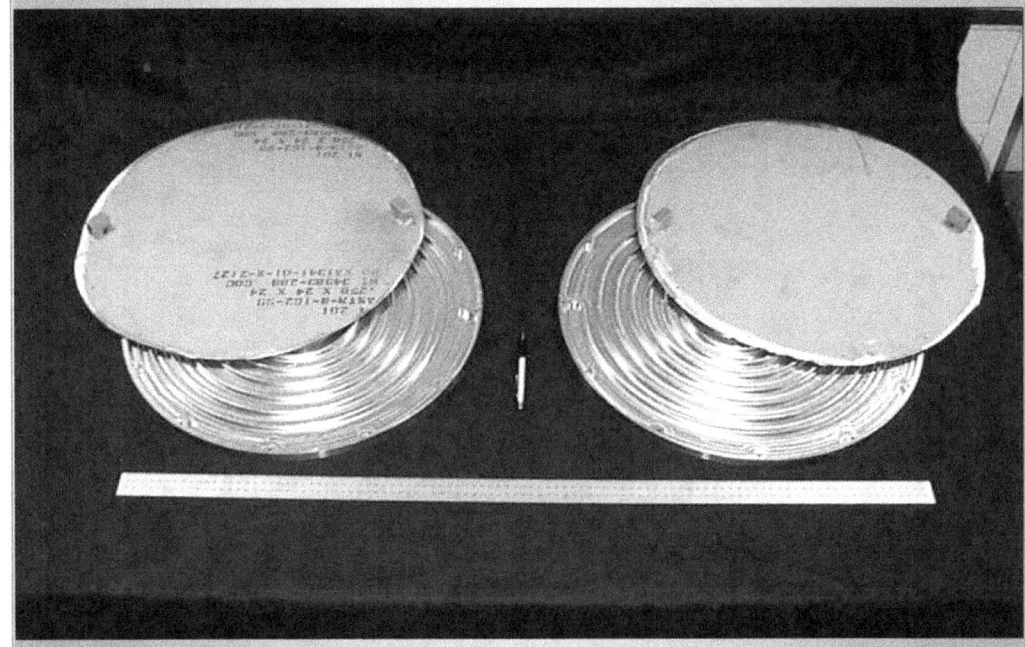

Coolant Plate (Spiral Channel HX)

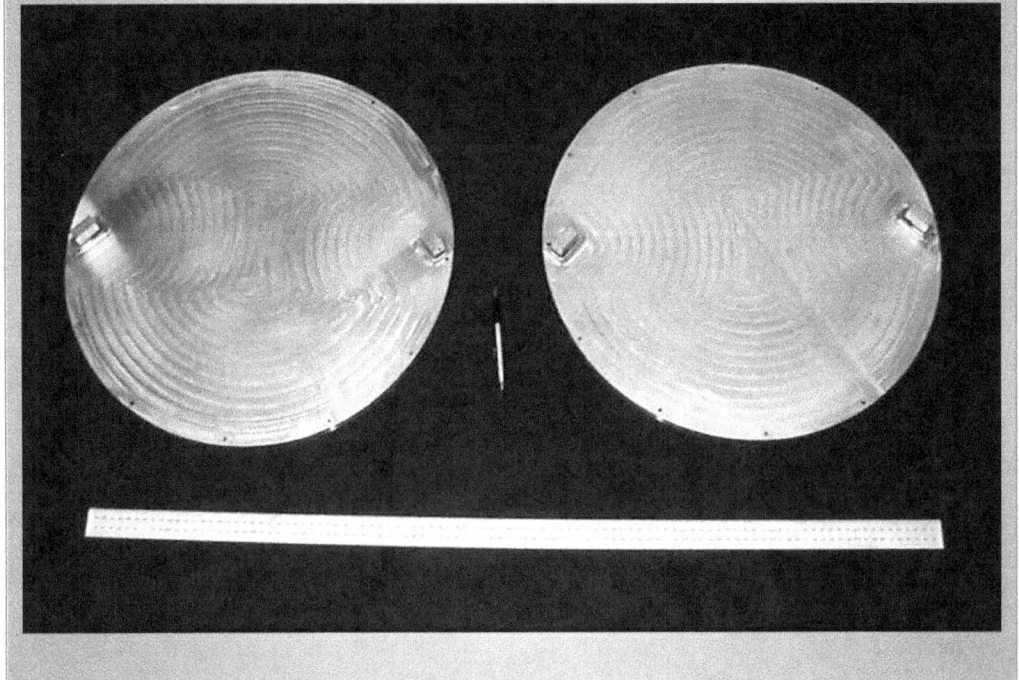

Coolant Plate (Spiral Channel HX)

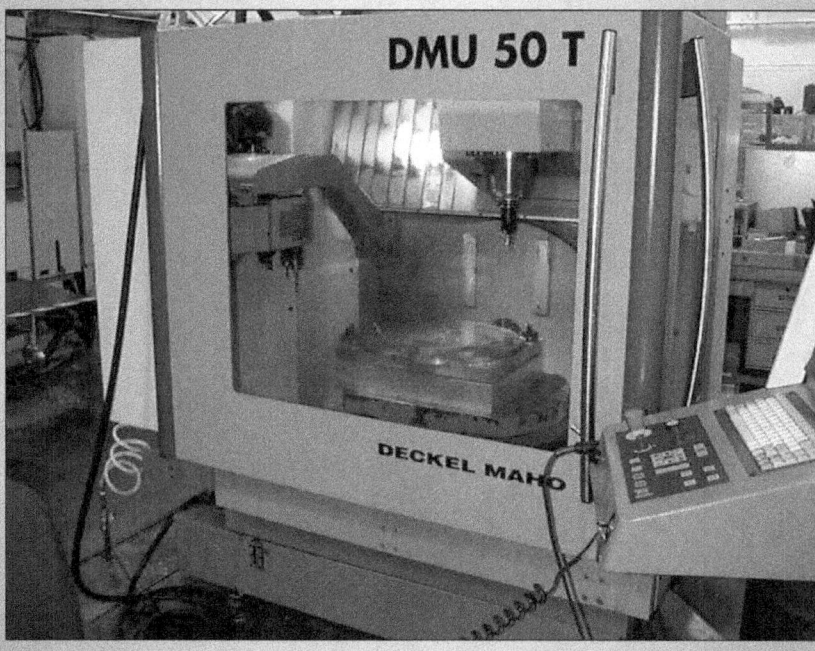

Coolant Plate (Spiral Channel HX)

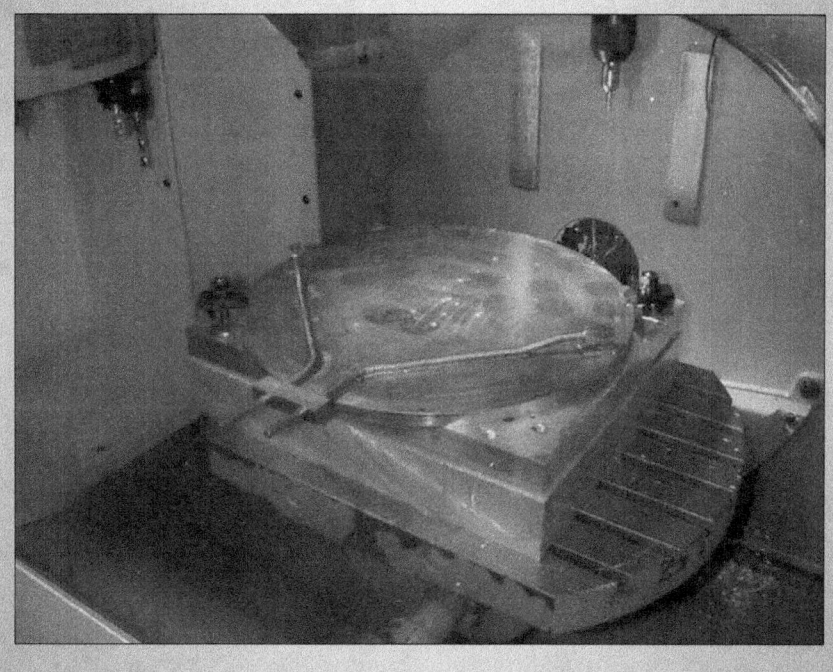

119

Edge Guard Assemblies –
Heater Ring, Coolant Ring, and
Water Jacket Ring

Fabrication, Vacuum Brazing

Edge Guard Rings – Initial Bending, Machining and Welding

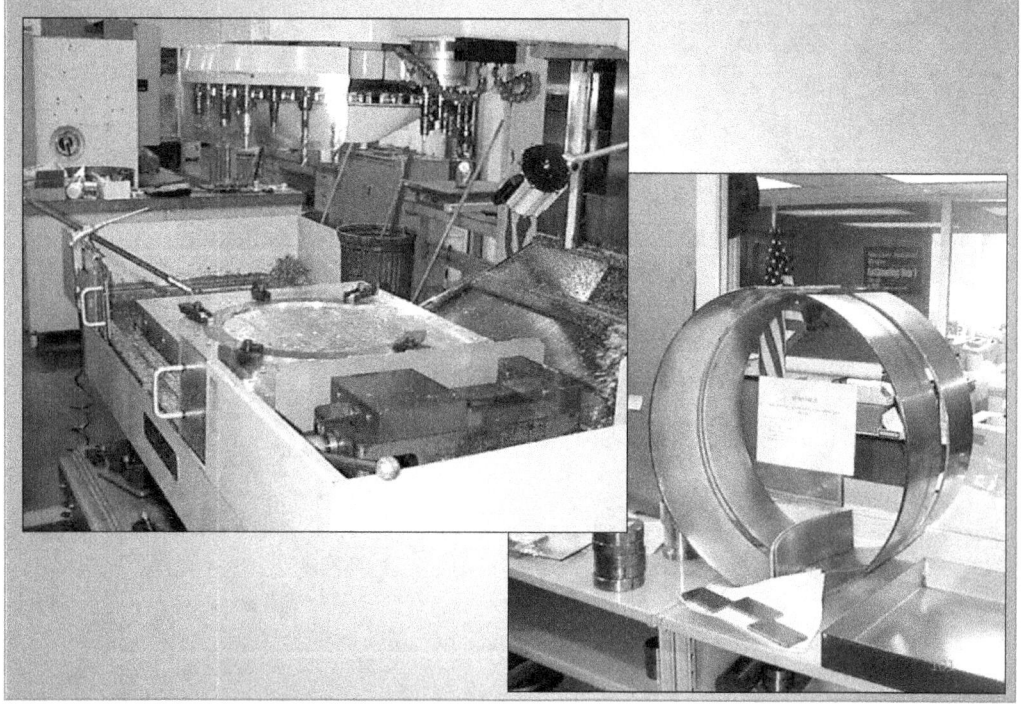

Edge Guard Rings – Initial Bending/Welding

Copper

Nickel 201

Edge Guard Rings – Initial Bending/Welding

Edge Guard Rings – Finish Lathe Cutting

Edge Guard Rings – Finish Lathe Cutting

Edge Guard Rings – Finish Lathe Cutting

Heater Rings

Edge Guard Heater Rings – Heater Grooves

Edge Guard Heater Rings – Thermocouple Grooves

Edge Guard Heater Rings – Vacuum Brazing

Edge Guard Heater Rings – Vacuum Brazing

Type N Thermocouples placed in grooves in ring inner diameter

Edge Guard Heater Rings – Vacuum Brazing

Heaters placed in grooves in ring outer diameter

Edge Guard Heater Rings – Vacuum Brazing

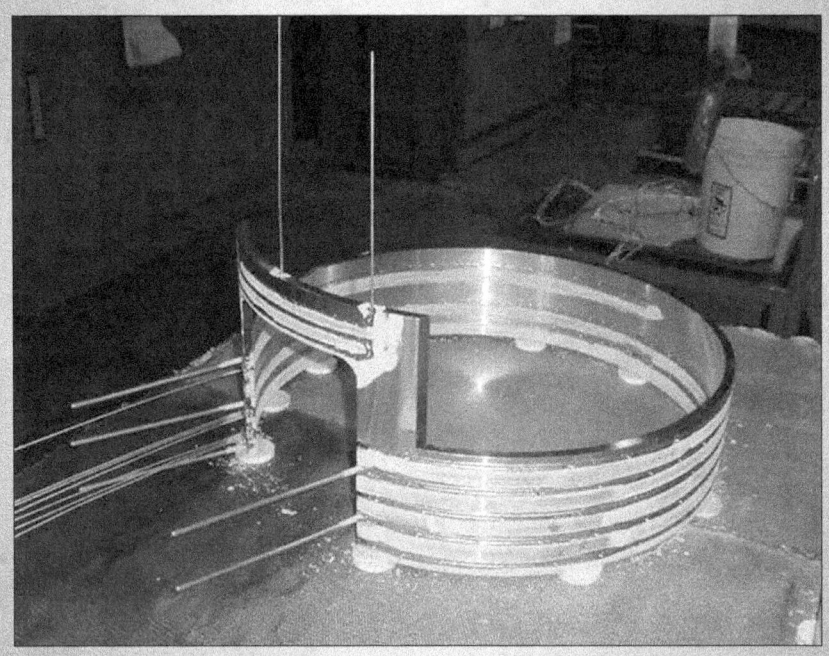

Edge Guard Heater Rings – Vacuum Brazing

Edge Guard Heater Rings – Vacuum Brazing

Edge Guard Heater Rings – Electrical Terminations

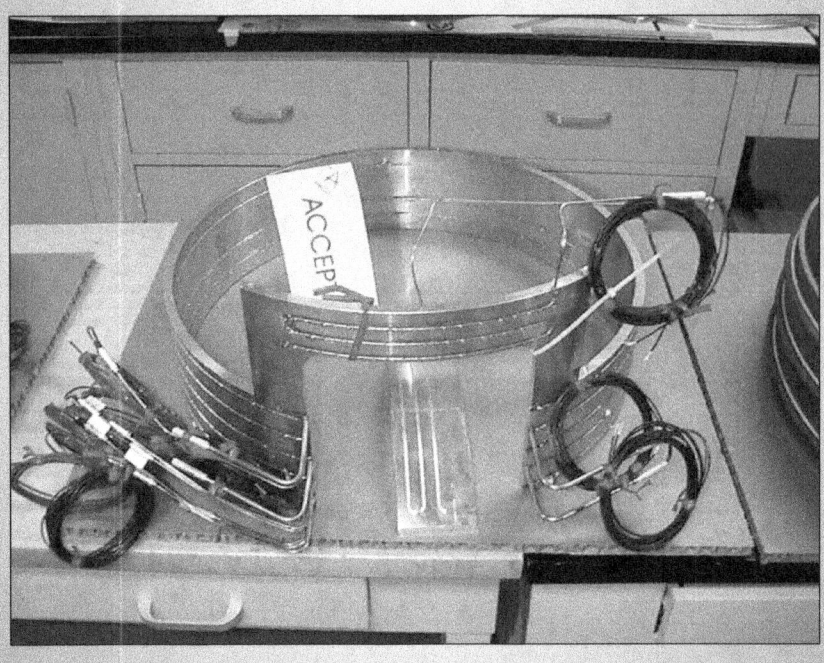

Edge Guard Heater Rings – Electrical Terminations

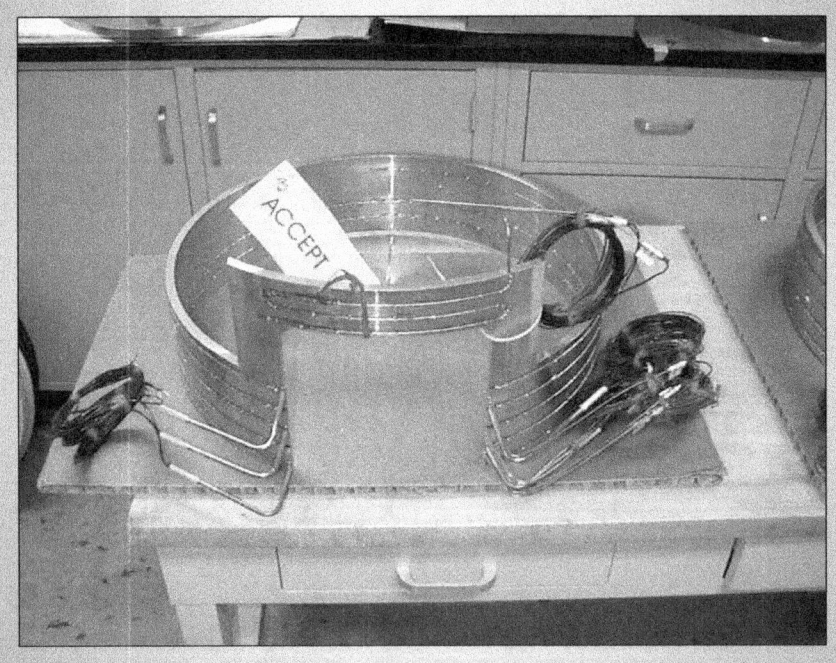

Coolant Rings

Edge Guard Coolant Rings – NIST Welding

Edge Guard Coolant Rings – NIST Welding

Water Jacket Rings

Edge Guard Coolant Rings – NIST Machining

Edge Guard Coolant Rings – NIST Machining

Assemble Edge Guard Rings

Assemble Large to Small Diameters
Flexible Thermal Insulation Between Rings

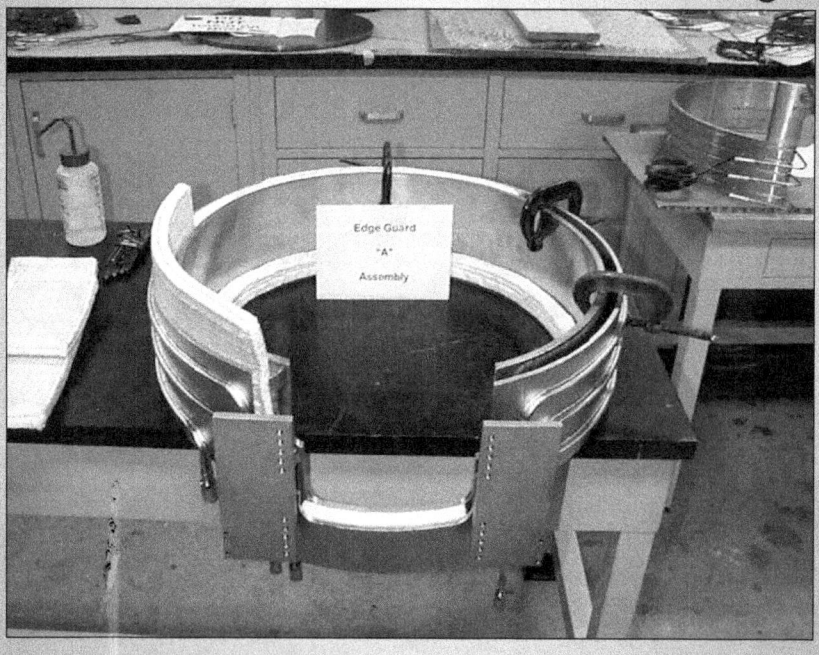

Assemble Large to Small Diameters
Flexible Thermal Insulation Between Rings

Assemble Large to Small Diameters
Flexible Thermal Insulation Between Rings

Assemble Large to Small Diameters
Flexible Thermal Insulation Between Rings

Hand-cut Flexible Thermal Insulation

Weight Check

Assemble Cold-Plate Components

Thermometer Plate – Opposite Surface

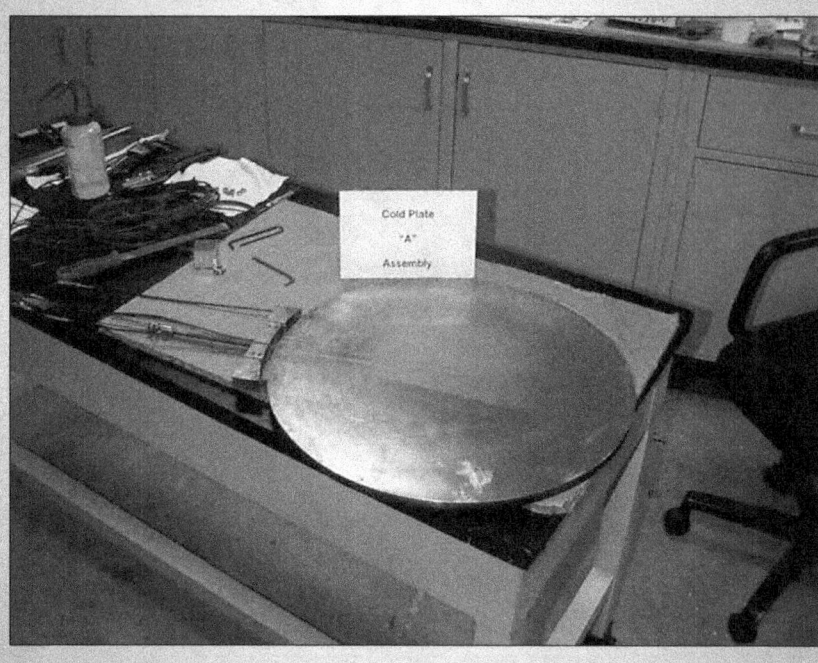

Thermometer Plate – Opposite Surface

Center area polished for thickness glass rod

Glass Cloth

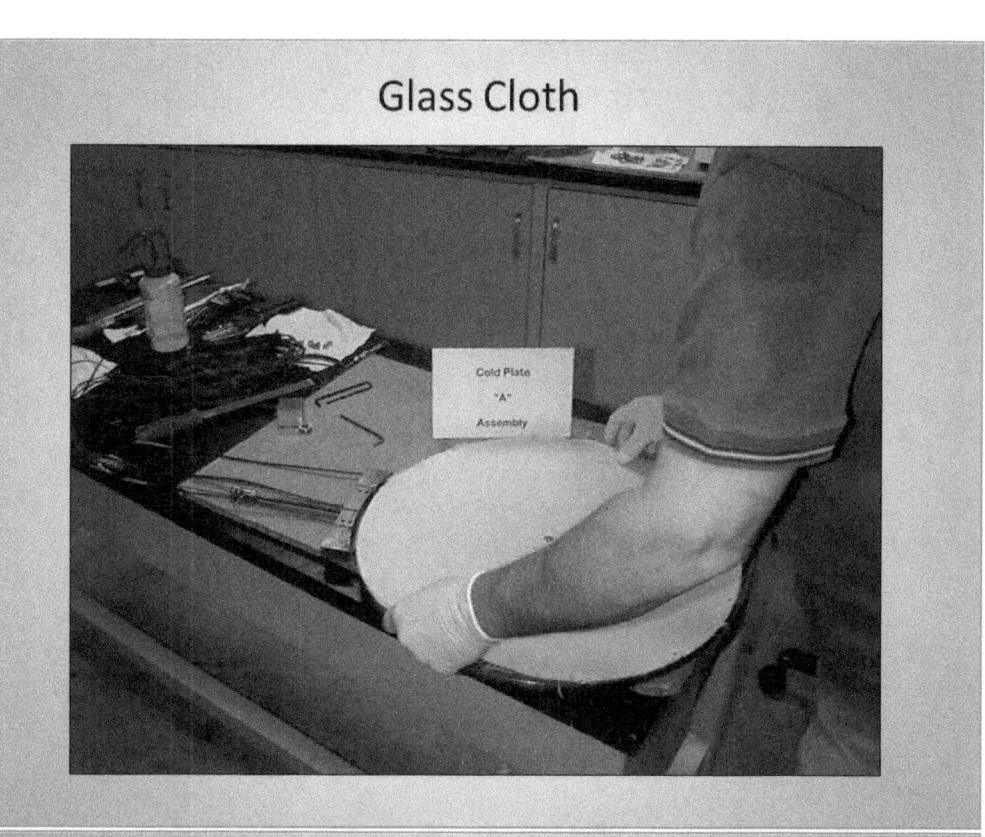

Install Heater Cold-Plate (Trim Glass Cloth)

Next Layer of Flexible Thermal Insulation

Install Coolant Plate

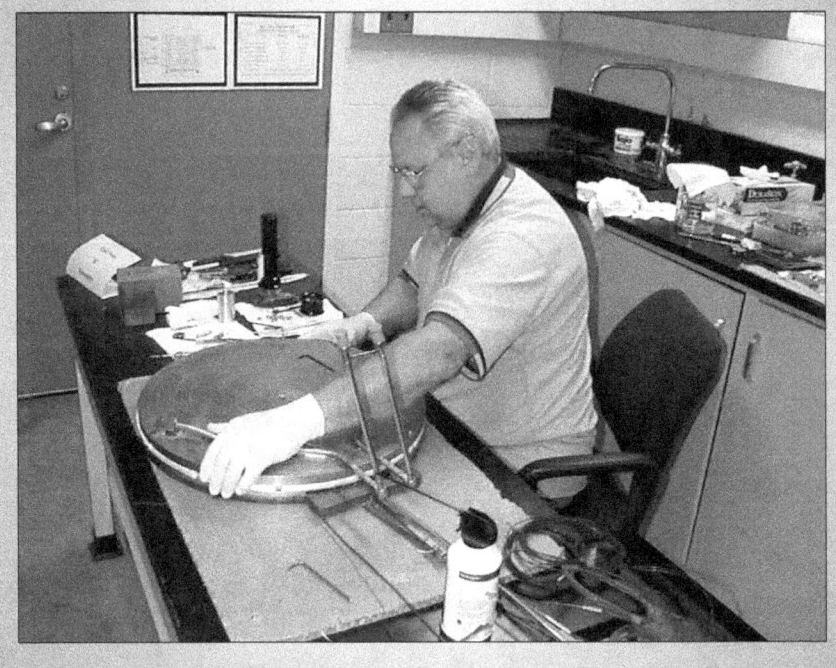

Install Rigid Insulation

Install Water Jacket

Install Flexible Wrap Around Water Jacket

Weight Check

Install Plates and Rings
on Overhead Rails

Initial Fit Check Edge Guard

Install Inboard Edge Guard
Check Overhead Rail Assembly

Install Inboard Edge Guard
Check Overhead Rail Assembly

Initial Fit Check Inboard Cold Plate Assembly

Install Inboard Cold Plate Assembly

Install Inboard Cold Plate Assembly

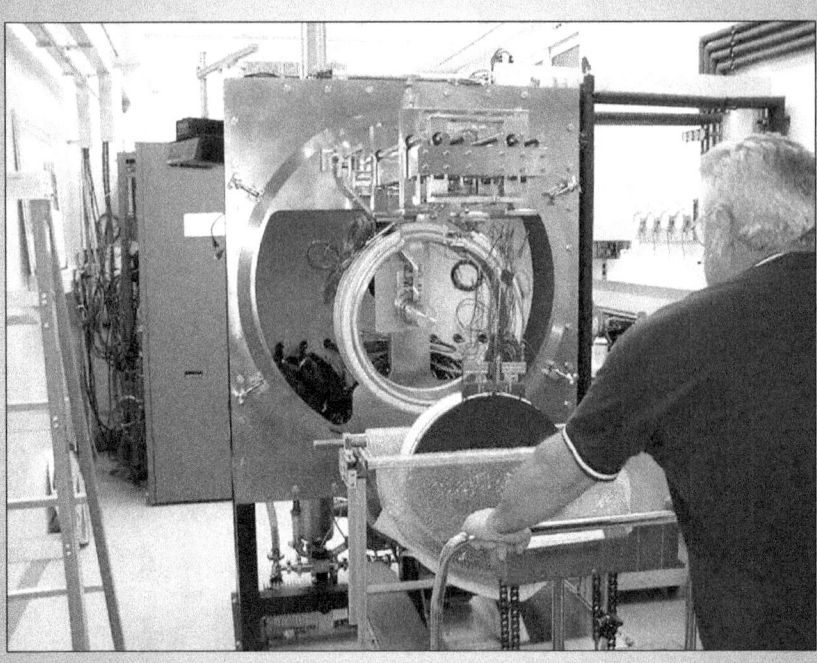

Install Inboard Cold Plate Assembly

Install Inboard Cold Plate Assembly

Install Inboard Cold Plate Assembly

Install Inboard Cold Plate Assembly

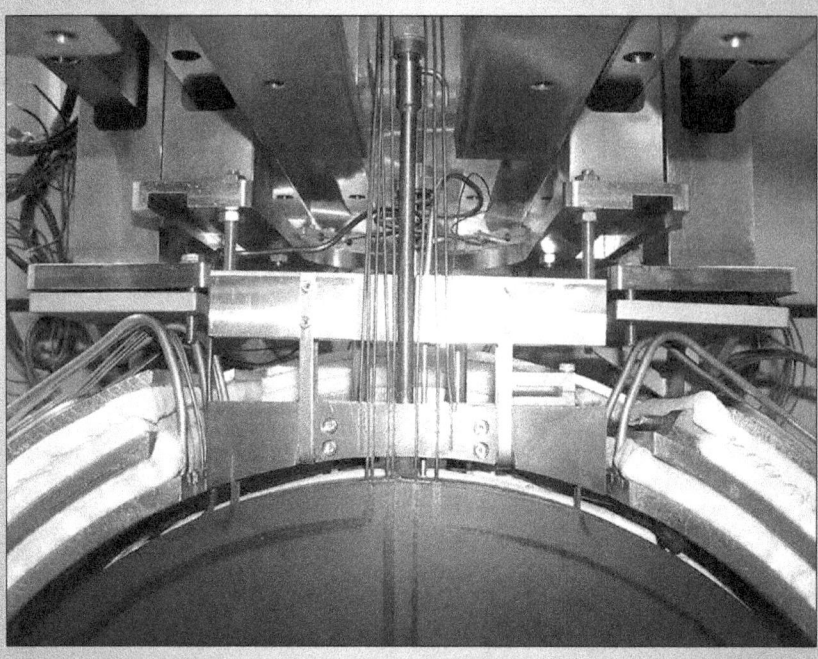

Install Inboard Cold Plate Assembly

Install Inboard Cold Plate Assembly

Install Outboard Cold Plate Assembly

Install Outboard Cold Plate Assembly

Install Outboard Cold Plate Assembly

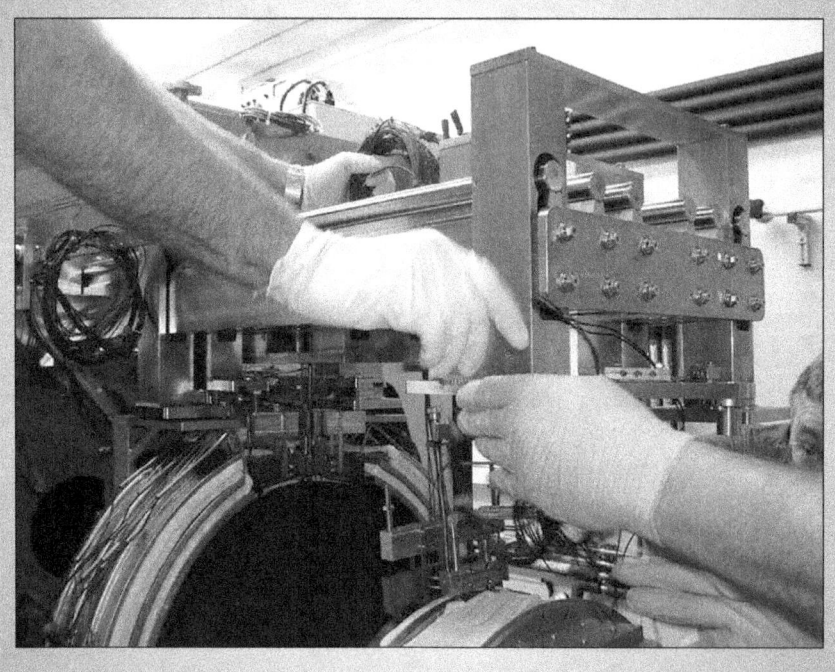

Install Outboard Cold Plate Assembly

Install Outboard Cold Plate Assembly

Install Outboard Edge Guard Assembly

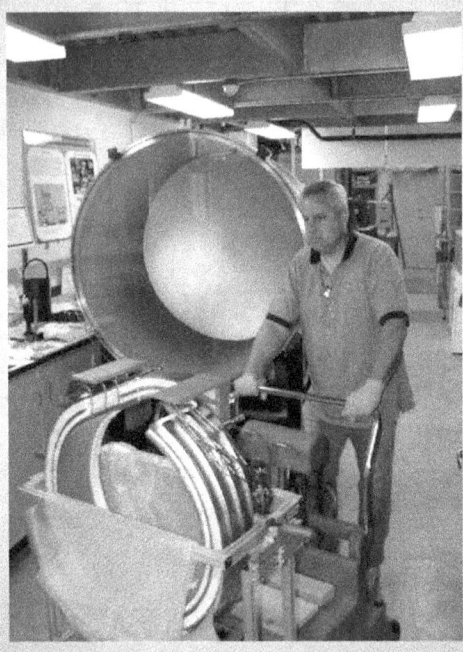

Install Outboard Edge Guard Assembly

Install Outboard Edge Guard Assembly

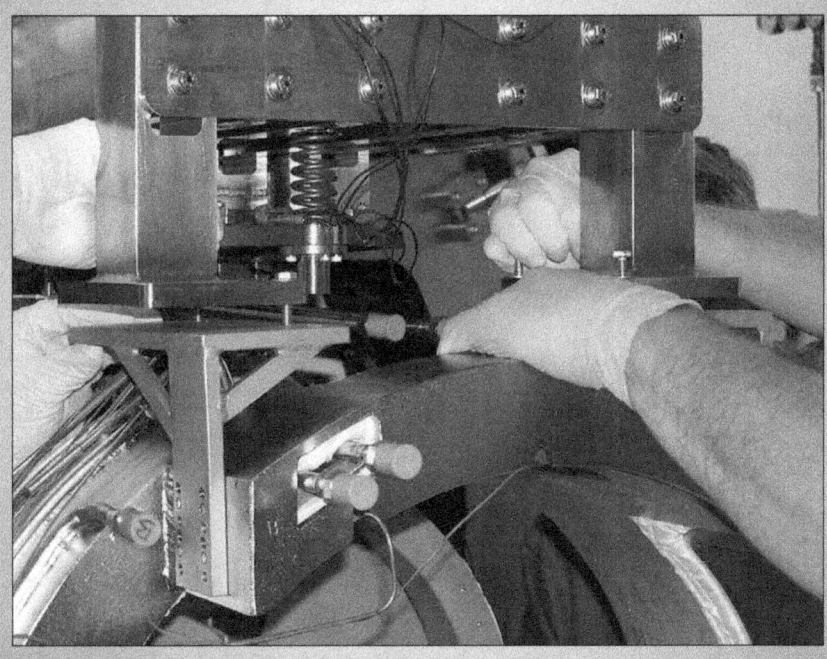

Install Outboard Edge Guard Assembly

Install Outboard Edge Guard Assembly

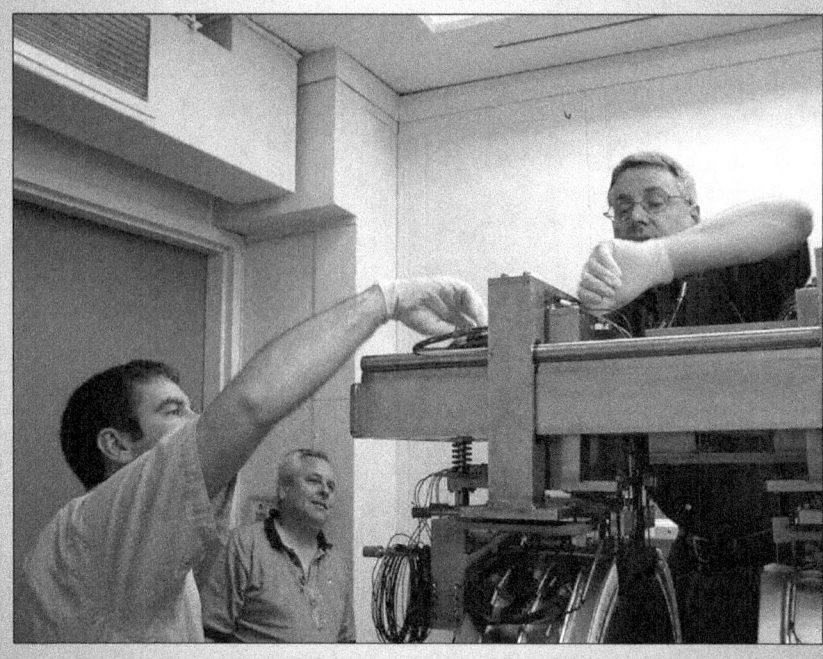

Plates and Rings Installed on Overhead Rails

Top View

Original Hot Plate Thermopile
Non-functional
(KP vs. 65 % Pd – 35 % Au)
Removal and Replacement

Prep for Sandblasting of Coating

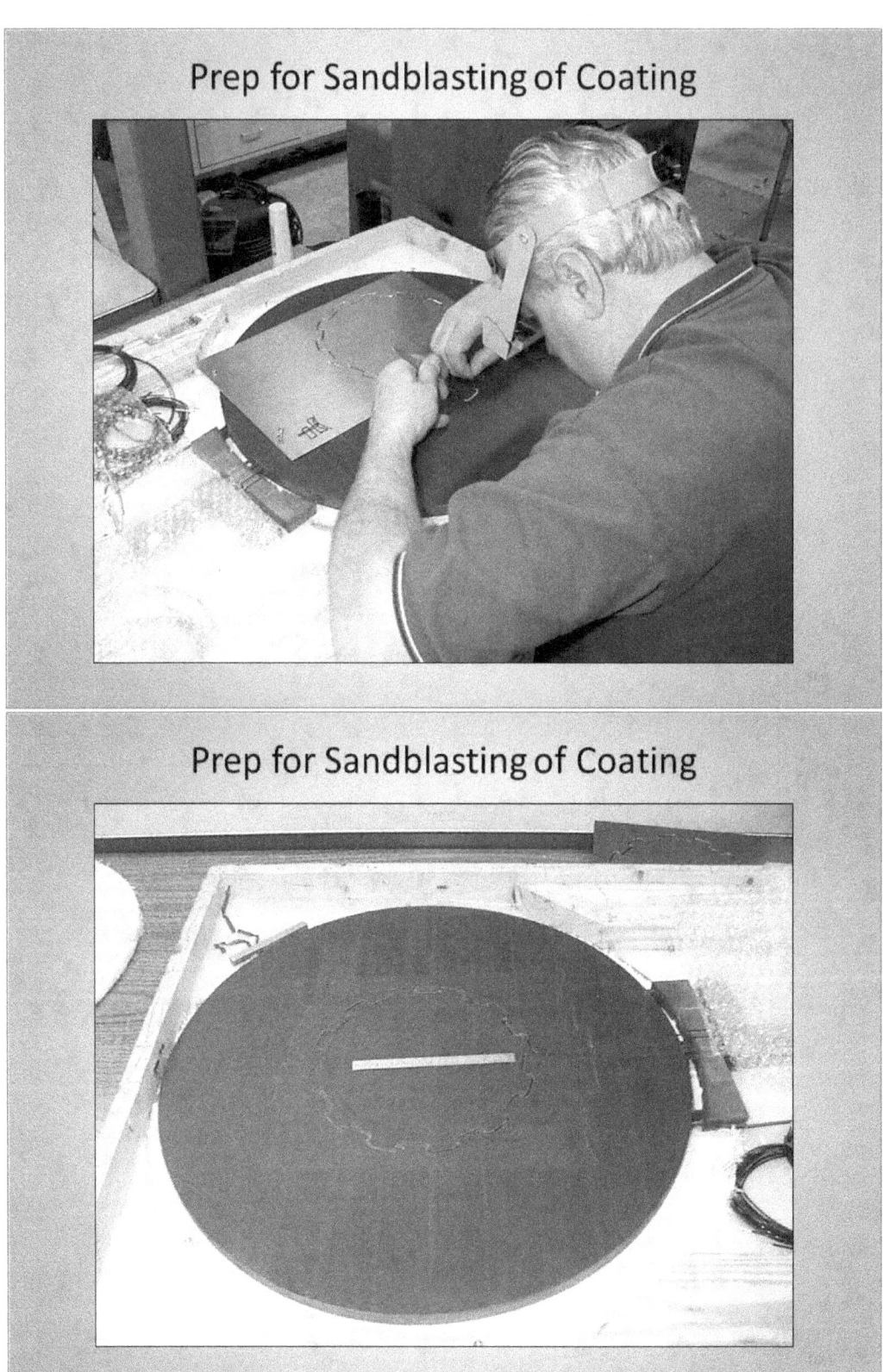

Prep for Sandblasting of Coating

Sandblasting Coating

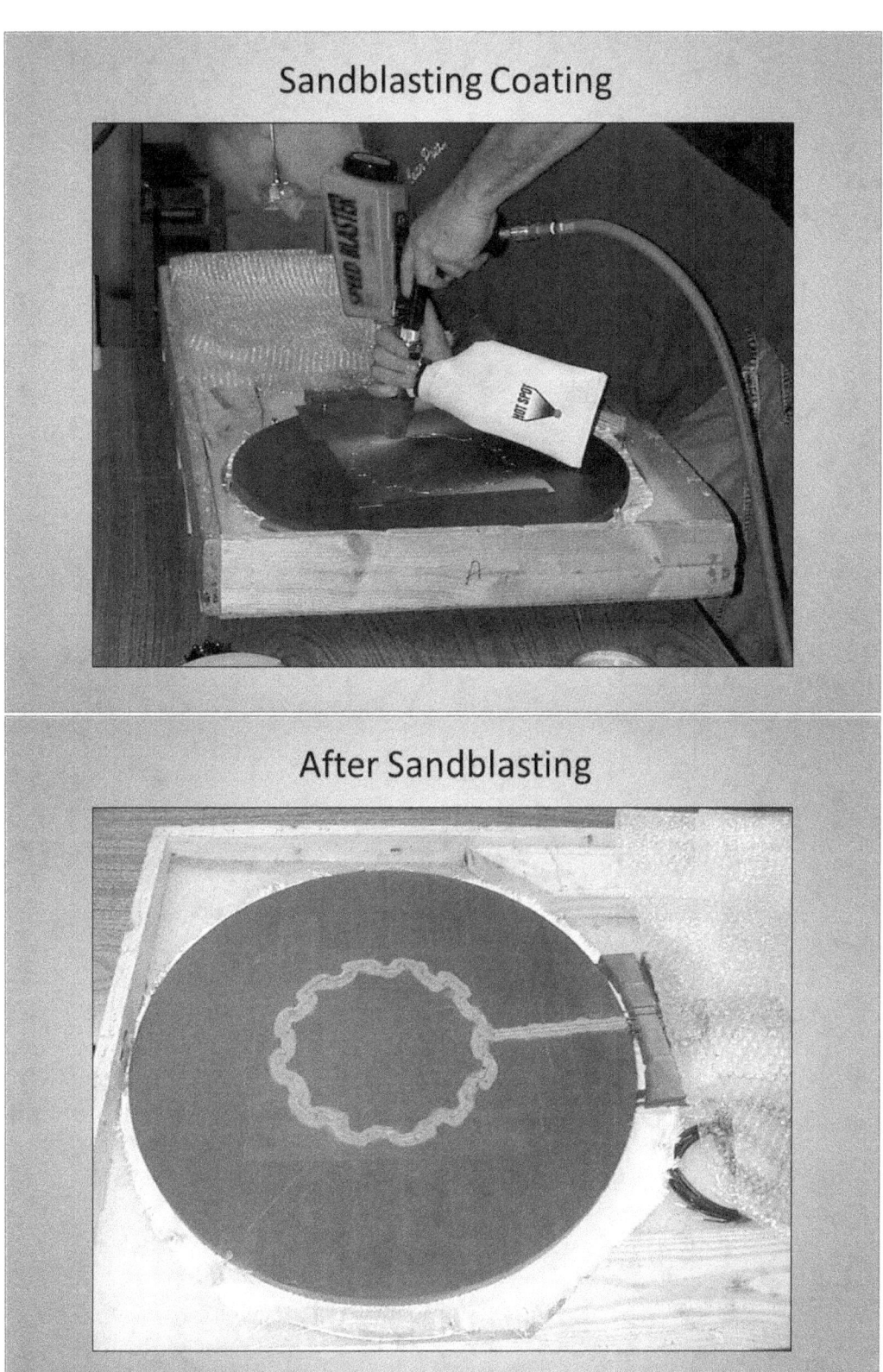

After Sandblasting

Tape Removal

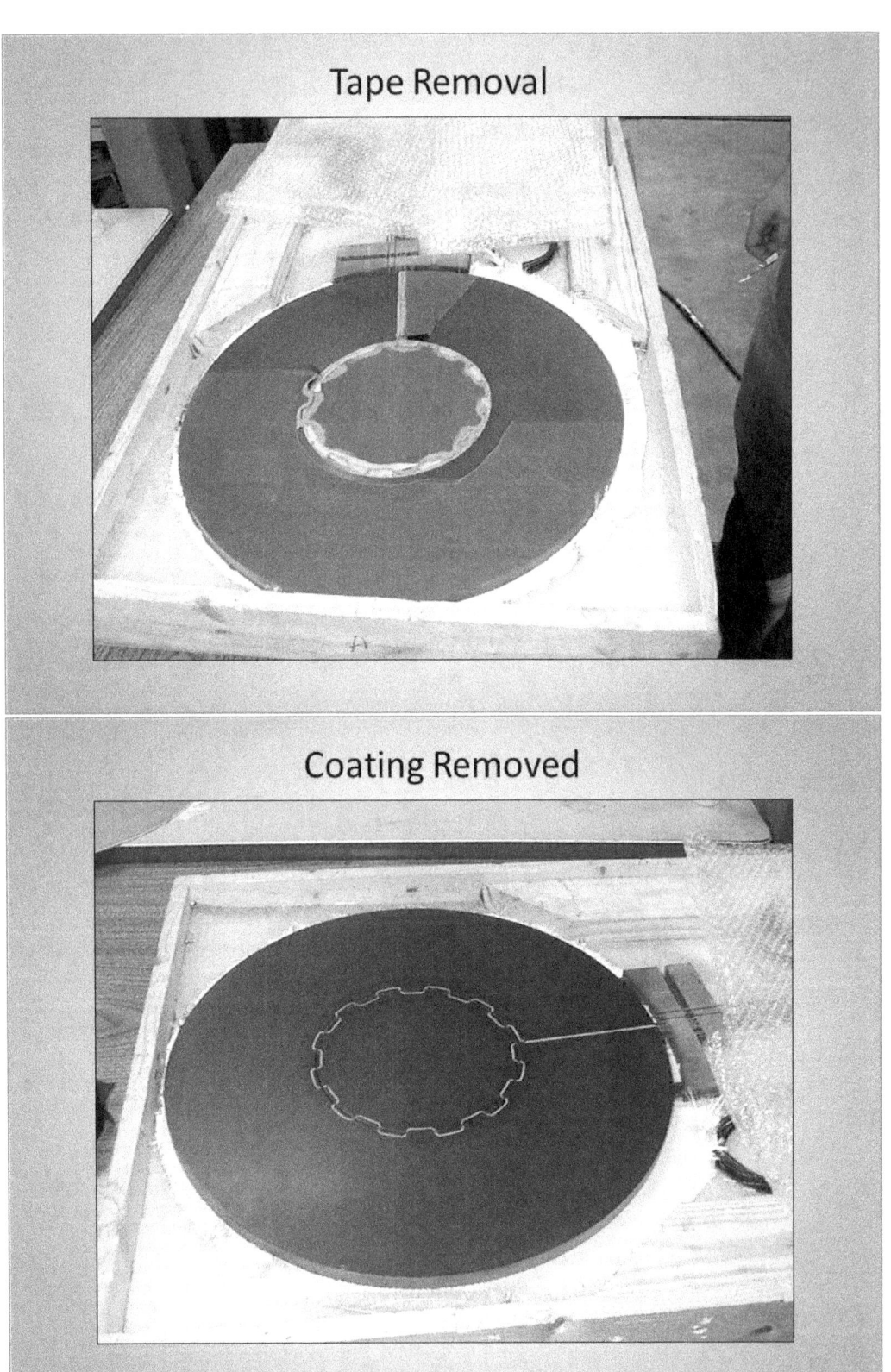

Coating Removed

Other Face

Other Face

Other Face

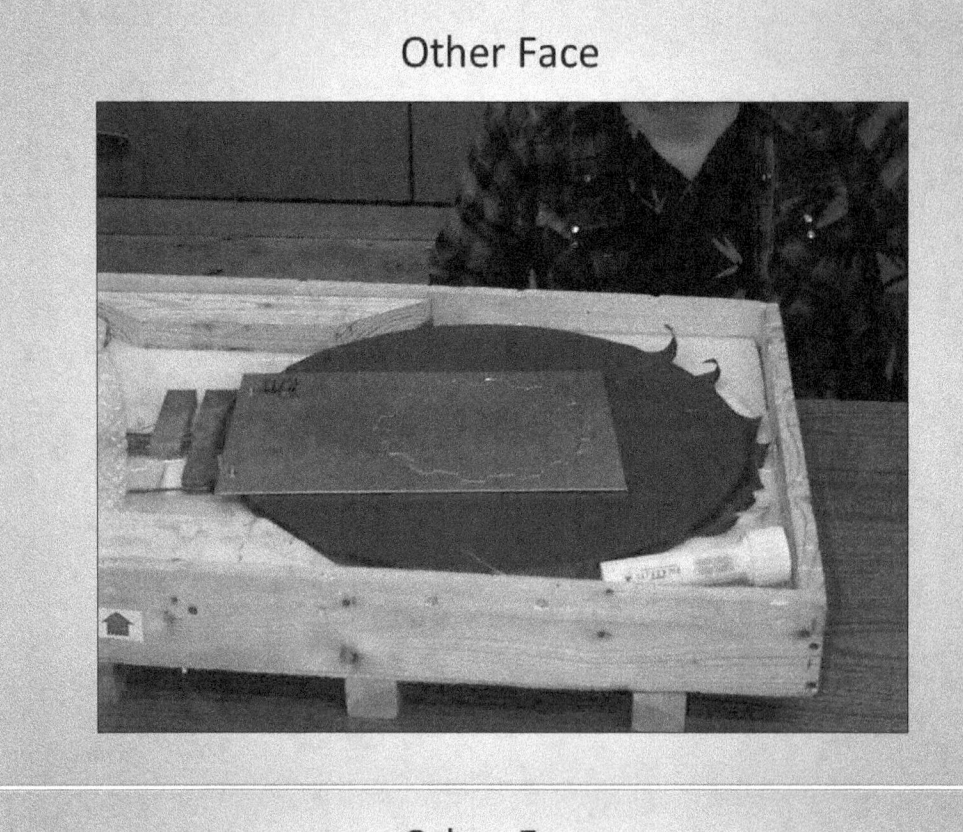

Other Face

Other Face

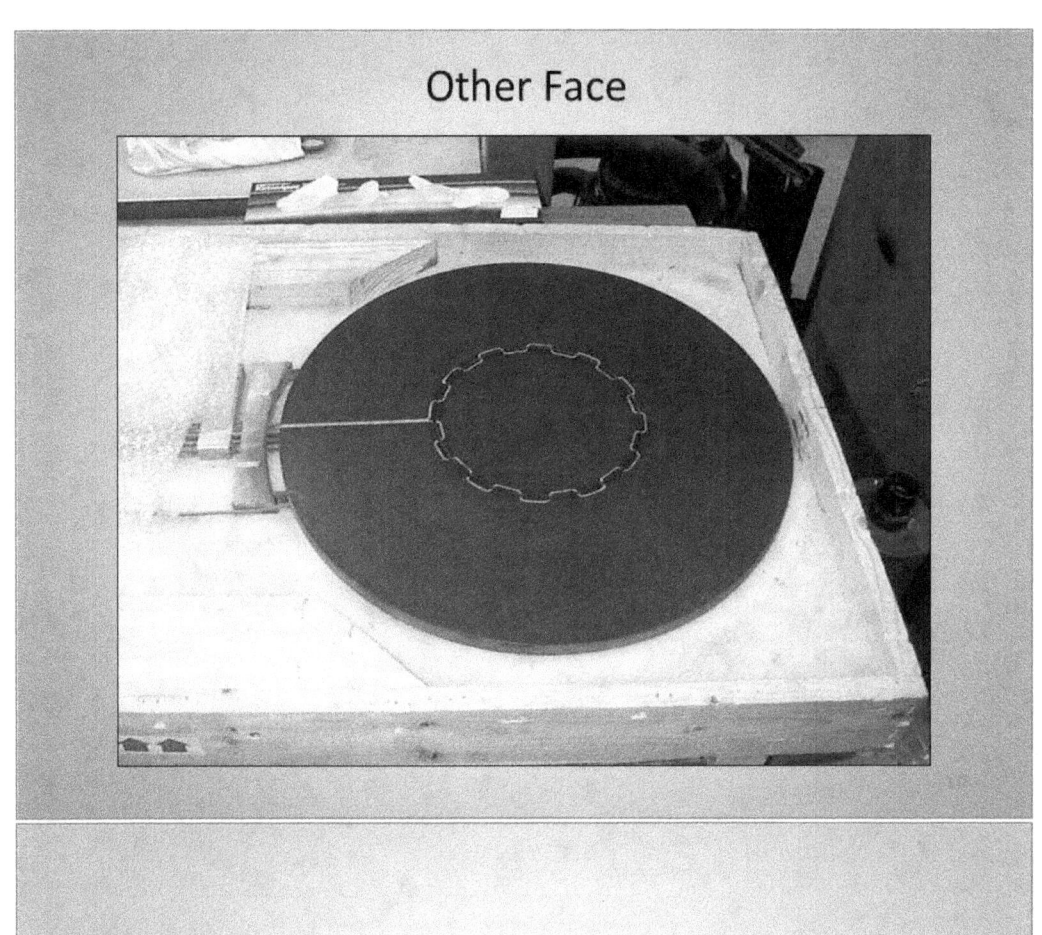

Machine out 1st Thermopile

Removal of Old Thermopile

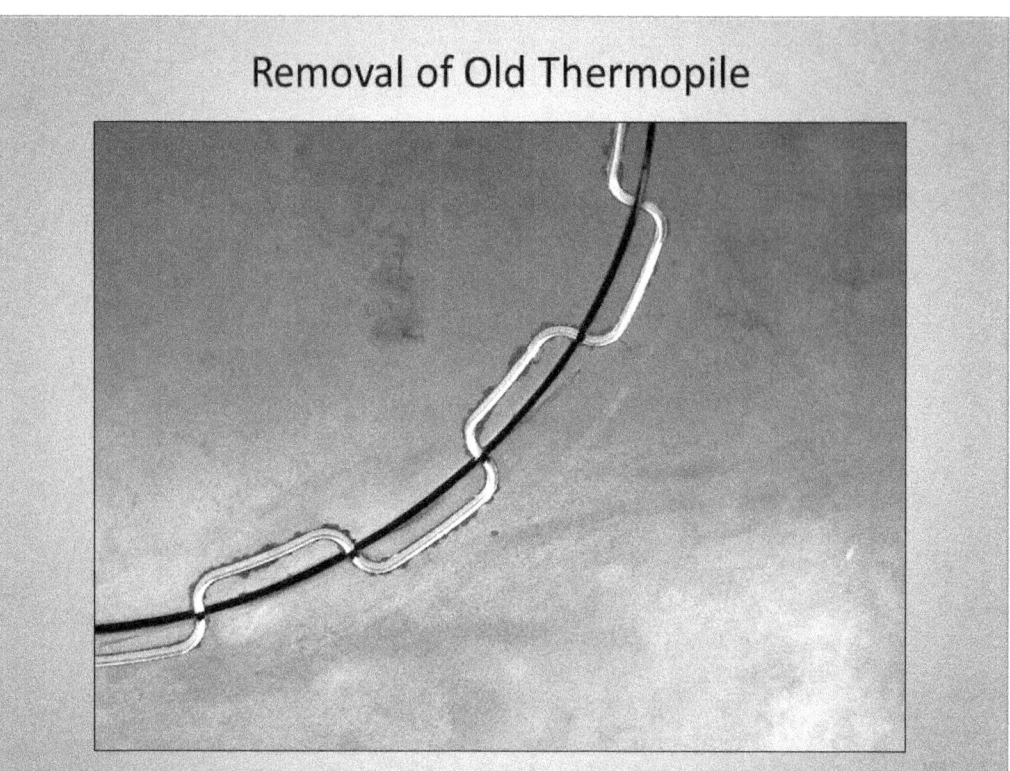

Installation of 2nd Hot Plate Thermopile (KP vs. KN)

Press-in New Thermopile

Press-in New Thermopile

Press-in New Thermopile

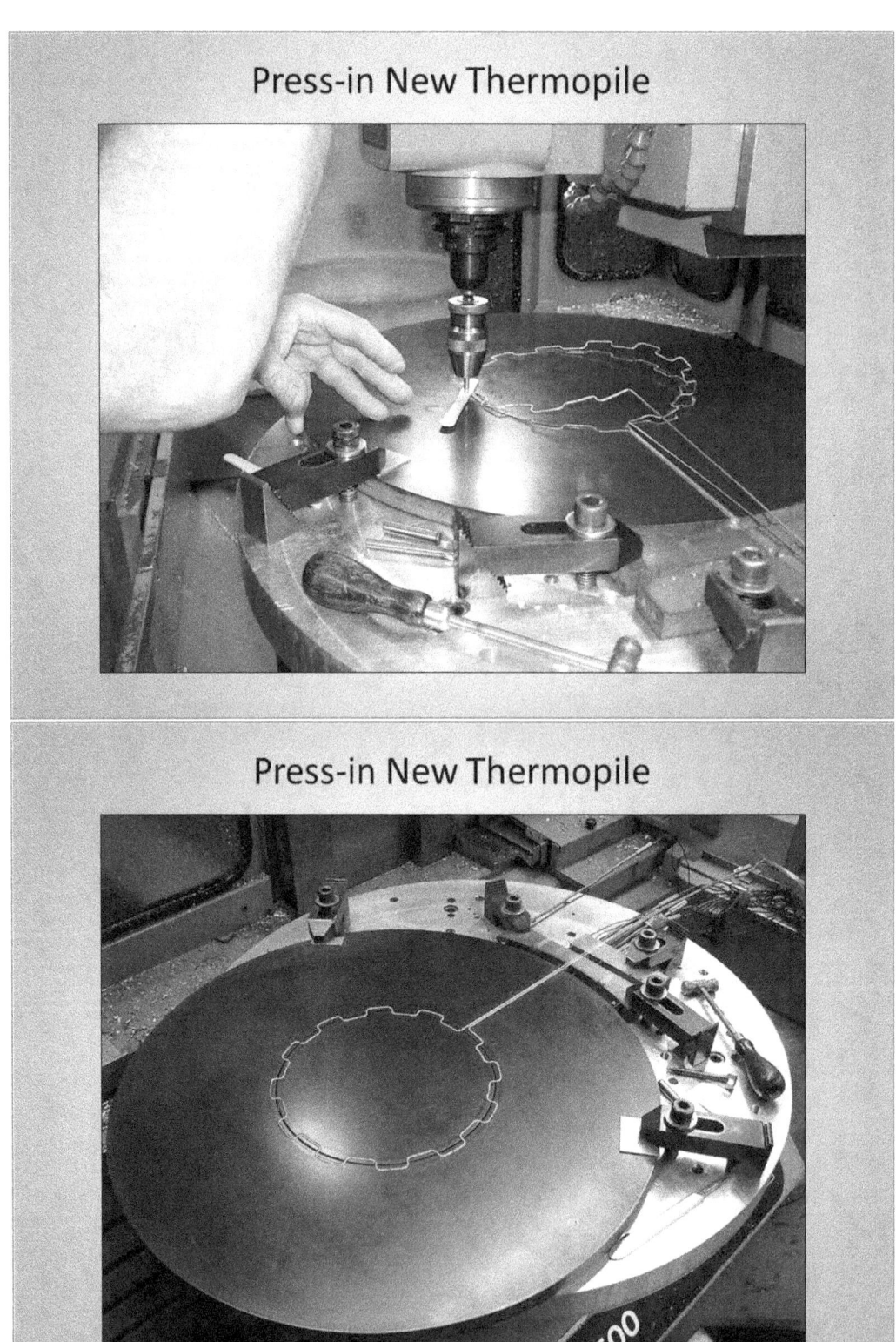

Press-in New Thermopile

Force Application and Thickness

Lower Support Assembly

Lower Support Assembly

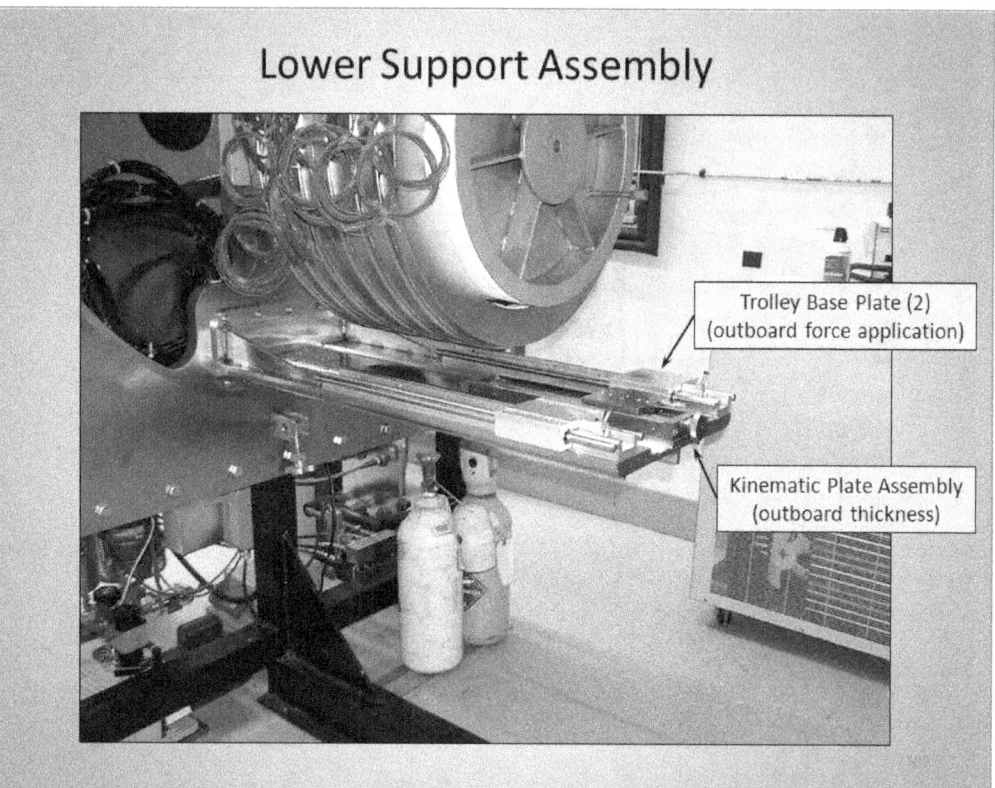

Trolley Base Plate (2)
(outboard force application)

Kinematic Plate Assembly
(outboard thickness)

Inboard Pedestal

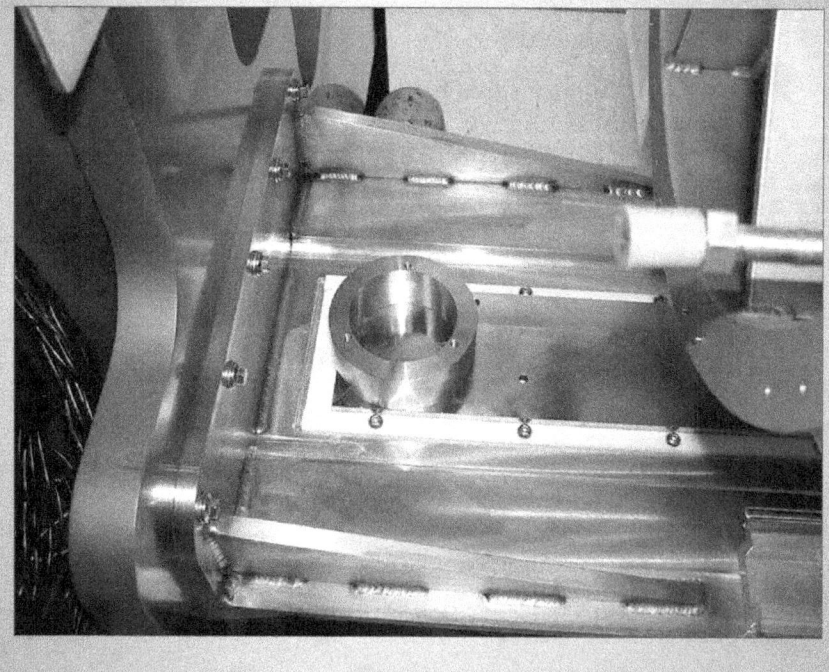

Kinematic Plate Assembly

Kinematic Mounts

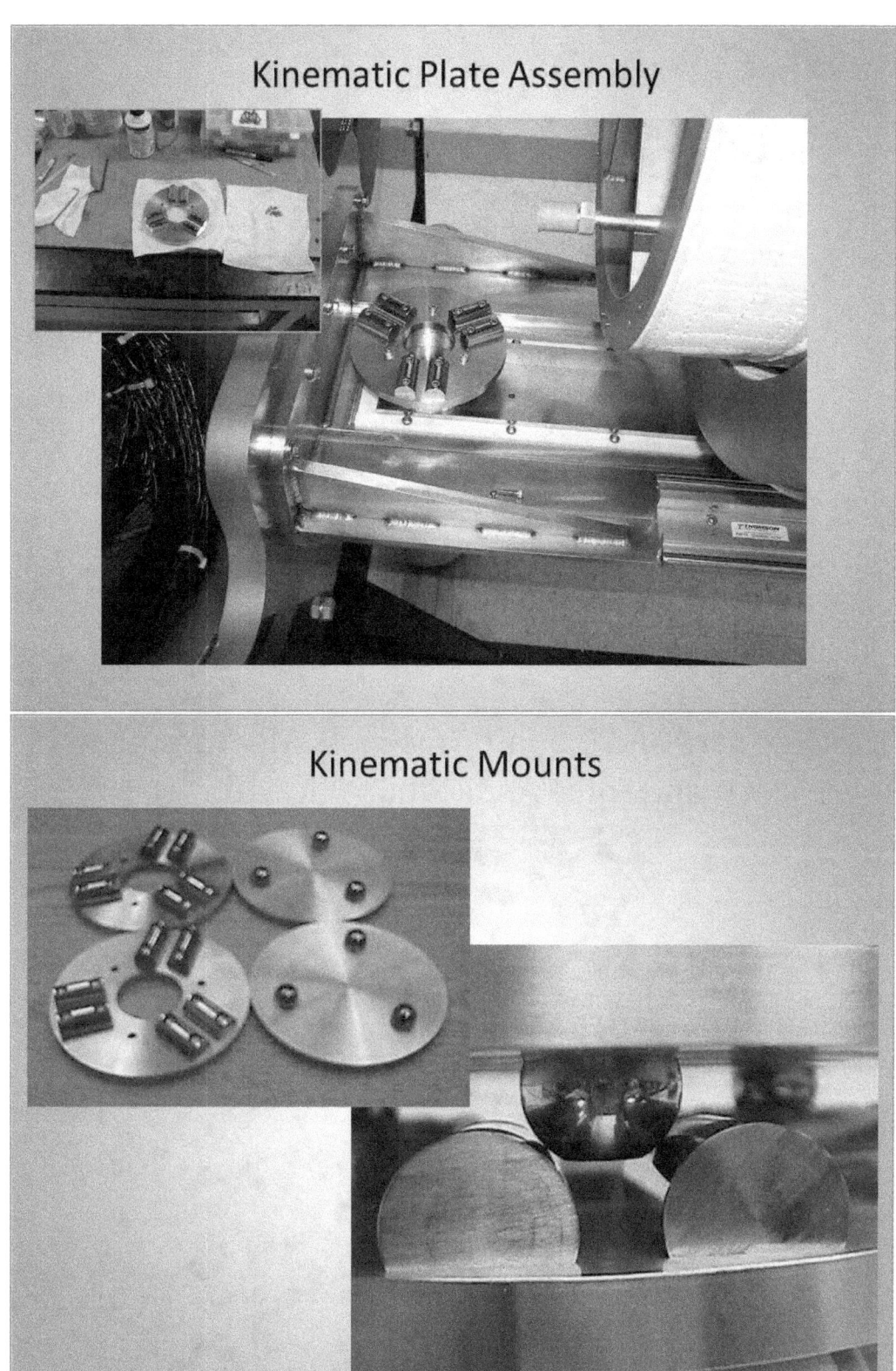

Inboard Stanchion Thickness Measurement

Outboard and Inboard Thickness Measurement and Force Application Backstop

Internal Bell Jar Wiring

Terminals (1 set)

Terminals for Thermocouple and Voltage

Isothermal Zone Box
for External Thermocouple Wiring

Pre-wire Terminals (copper telephone wire)

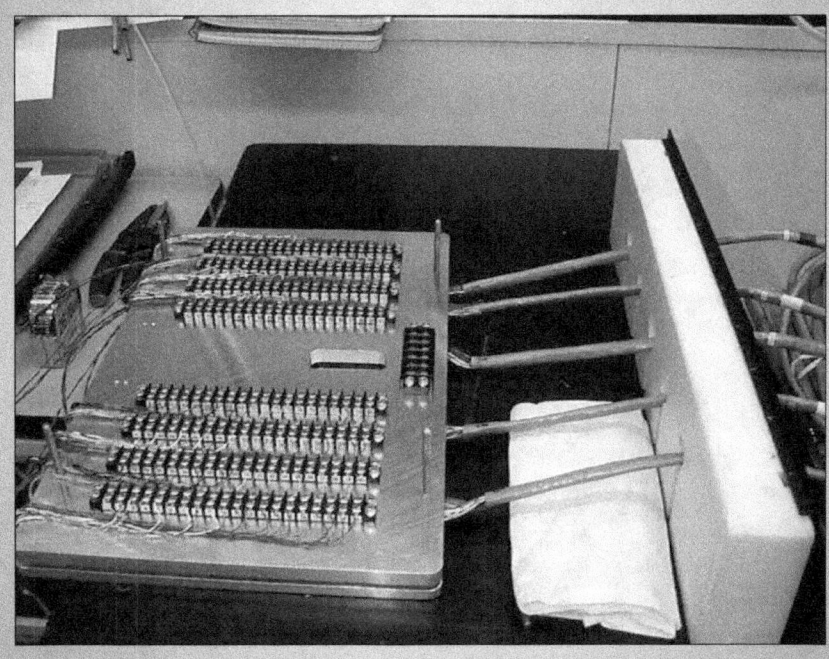

Instrument Panel (Retracted)

Install Terminals (copper telephone wire)

Install Thermocouple Wire (Bell Jar Feedthroughs)

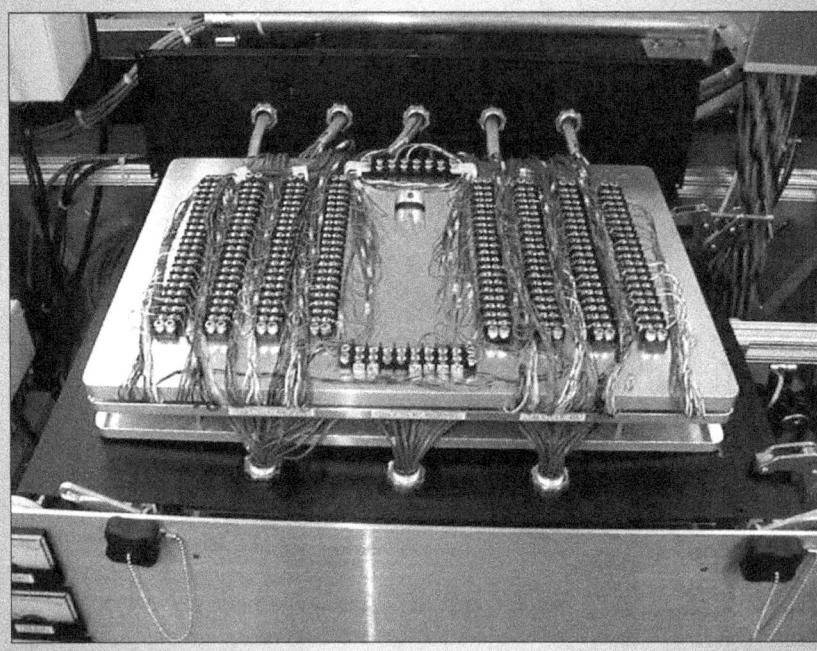

Install Lower Portion of Isothermal Box

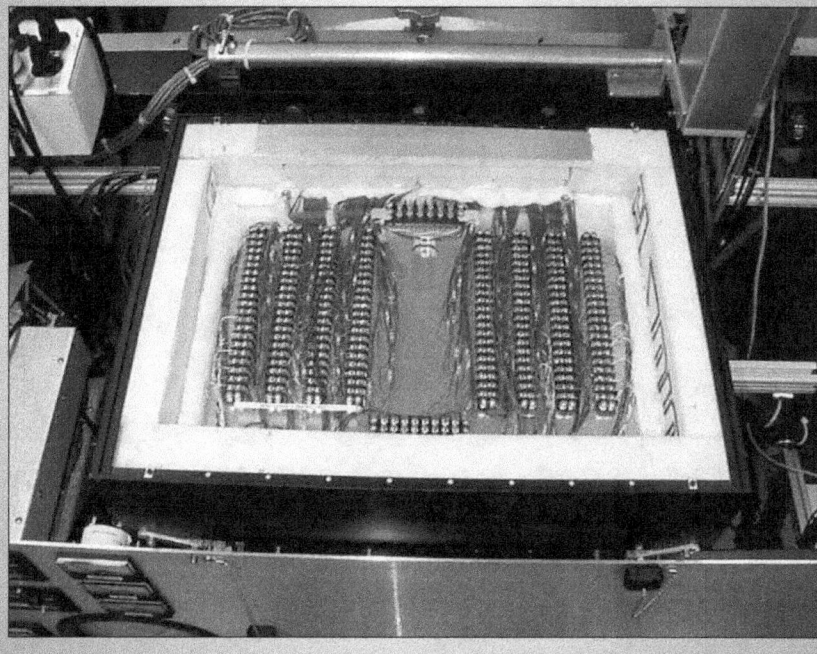

Insulate Top

Cover

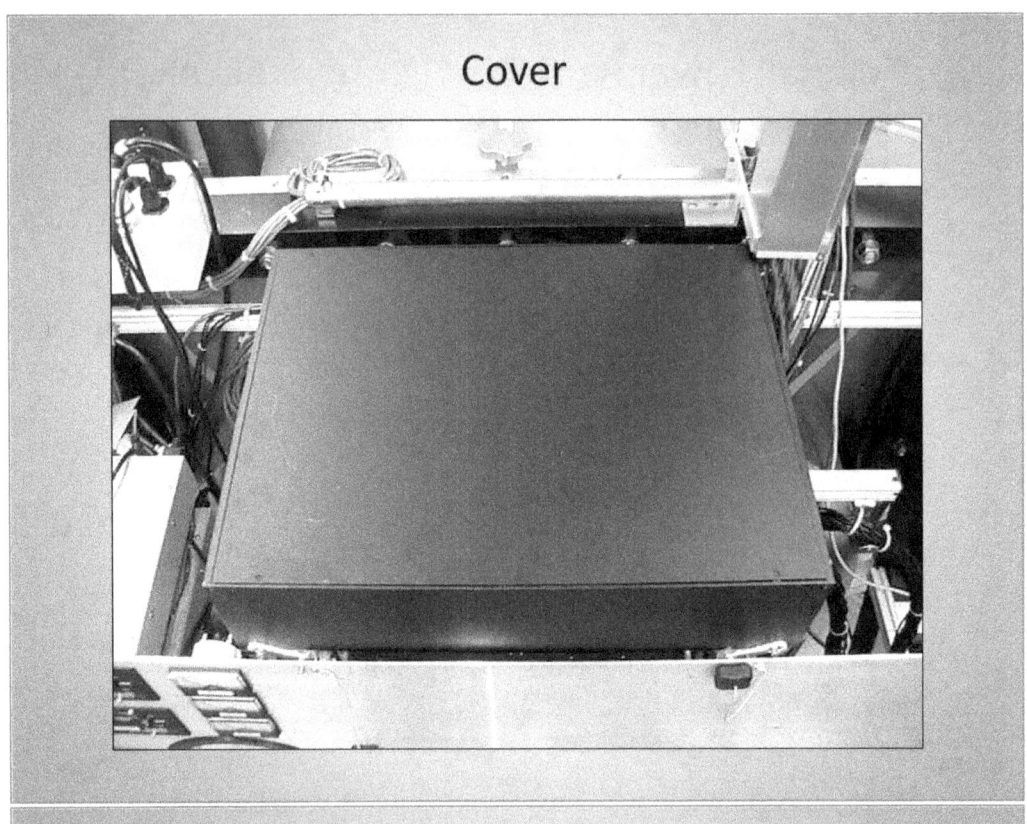

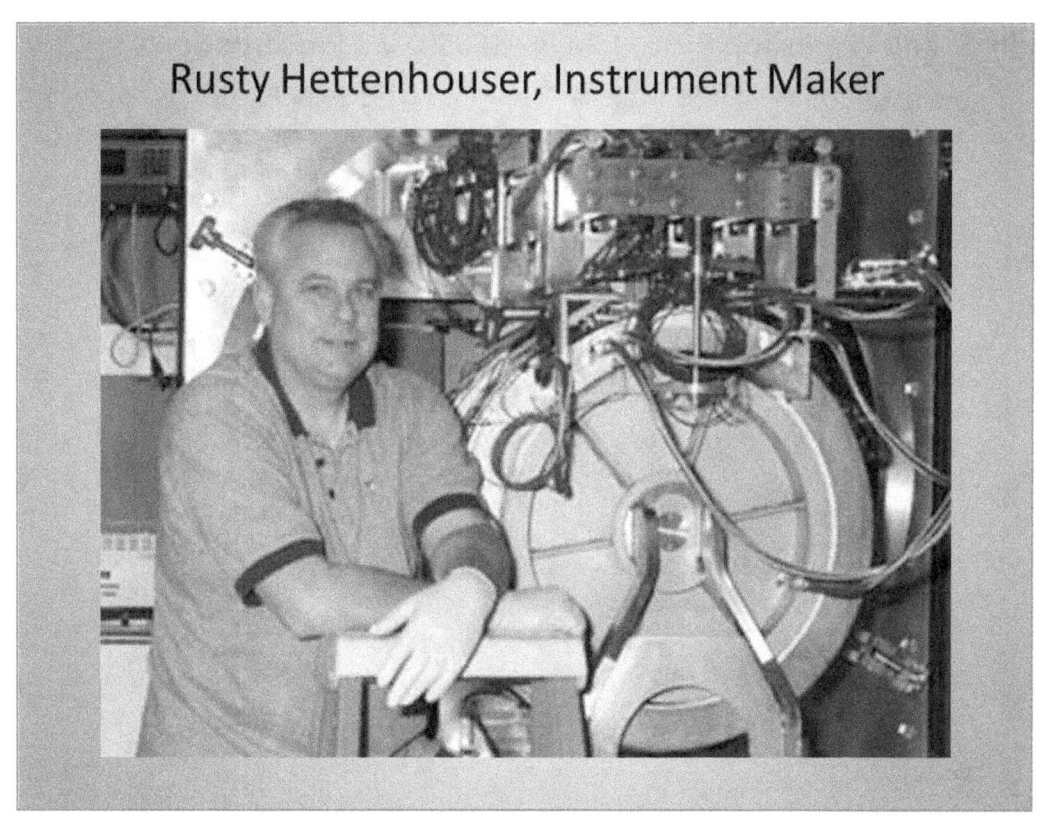

Rusty Hettenhouser, Instrument Maker

Simulation of PID Control for Guarded-Hot-Plate Apparatus

William C. Thomas
Professor Emeritus Mechanical Engineering
Virginia Polytechnic Institute & State University
Blacksburg, Virginia

Main Objective

Minimize Traditional Trial & Error Tuning Operations by Simulating Response with a Mathematical Model

- Reduce Time and Expense
- Improved Set of Controller Gains

Scope

- Describe Simulated Control Method
  - Mathematical Model Basis
    - Emphasis on the (Less Complex) 1-m GHP
  - Tuning Procedure for PID Gain Settings
- Typical Response Results
- Modeling Details

NIST 1-m Guarded-Hot-Plate

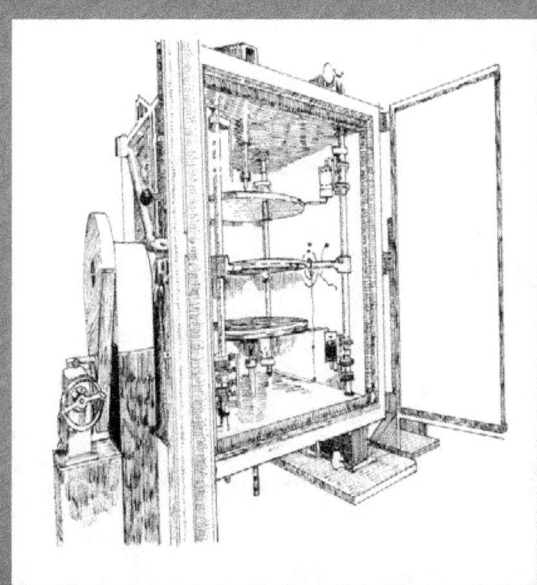

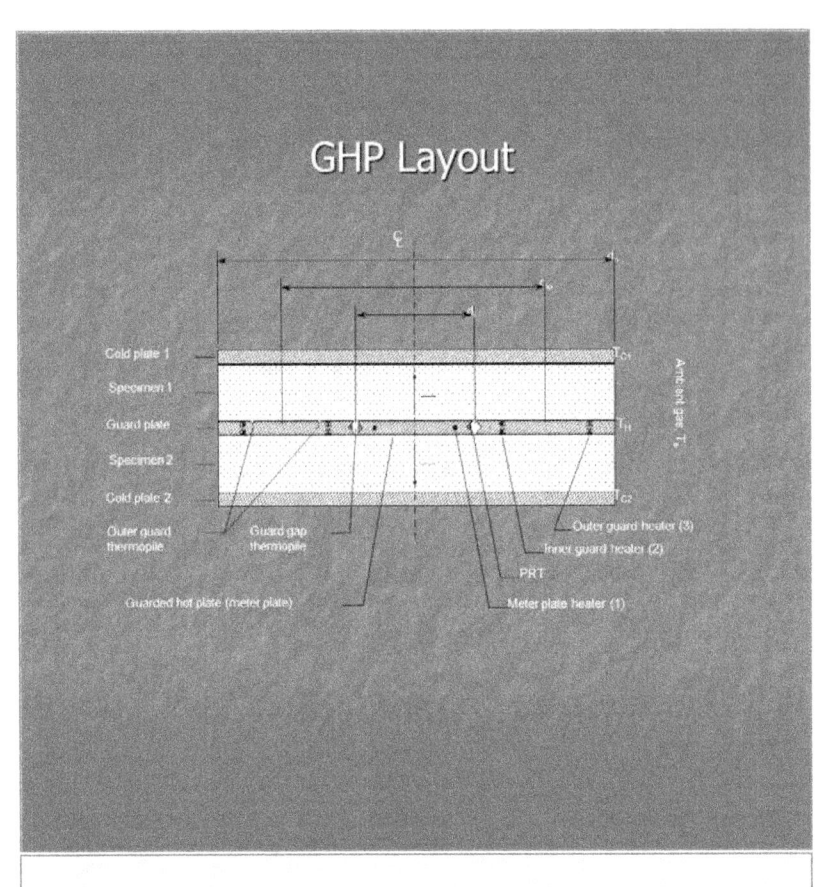

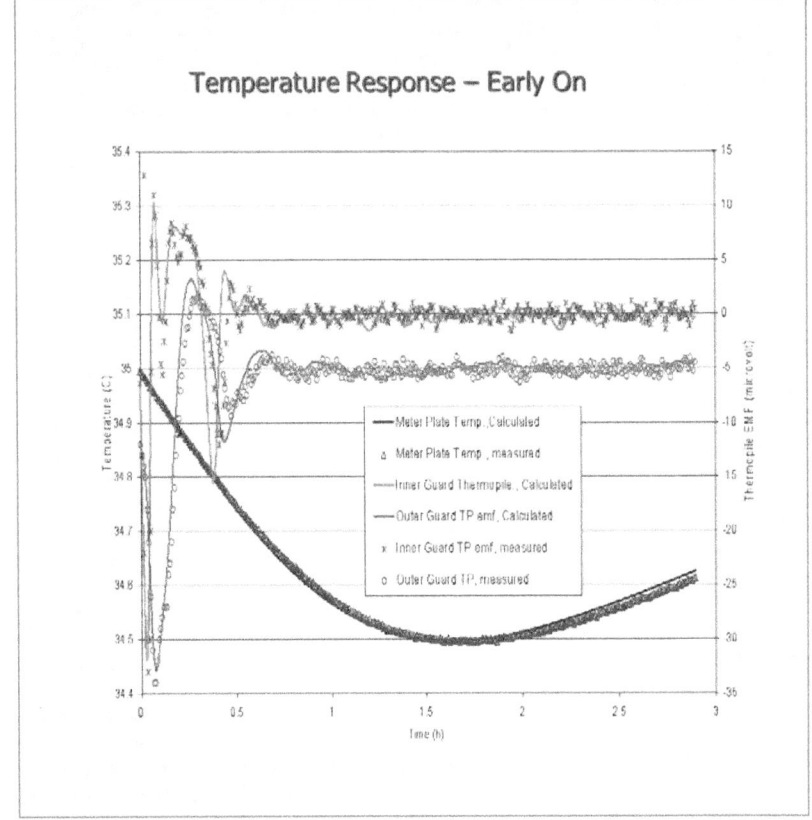

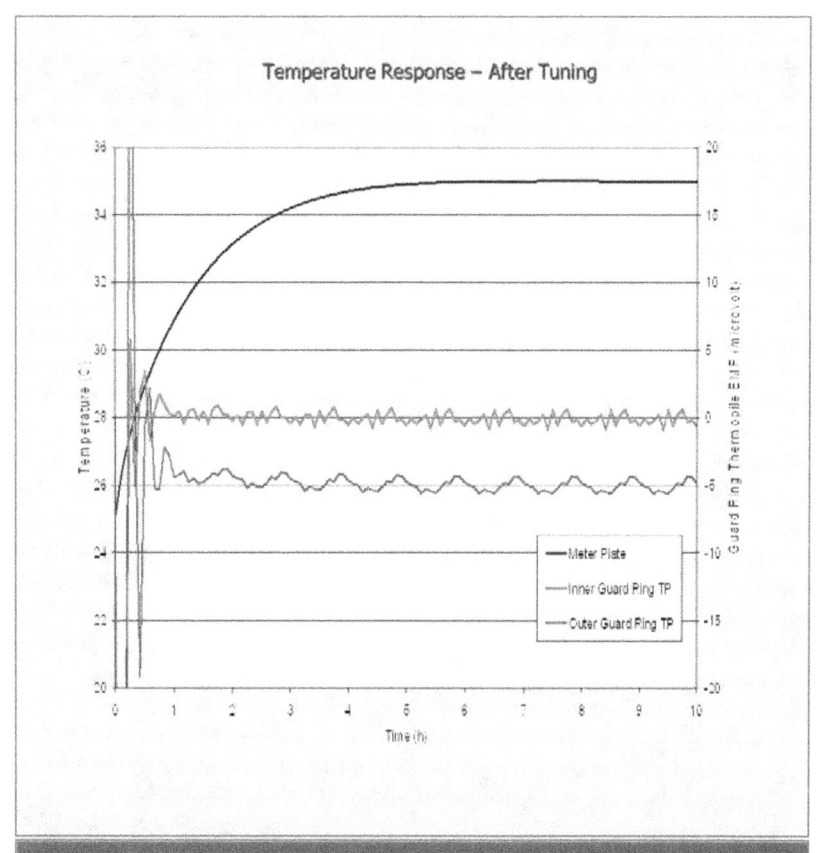

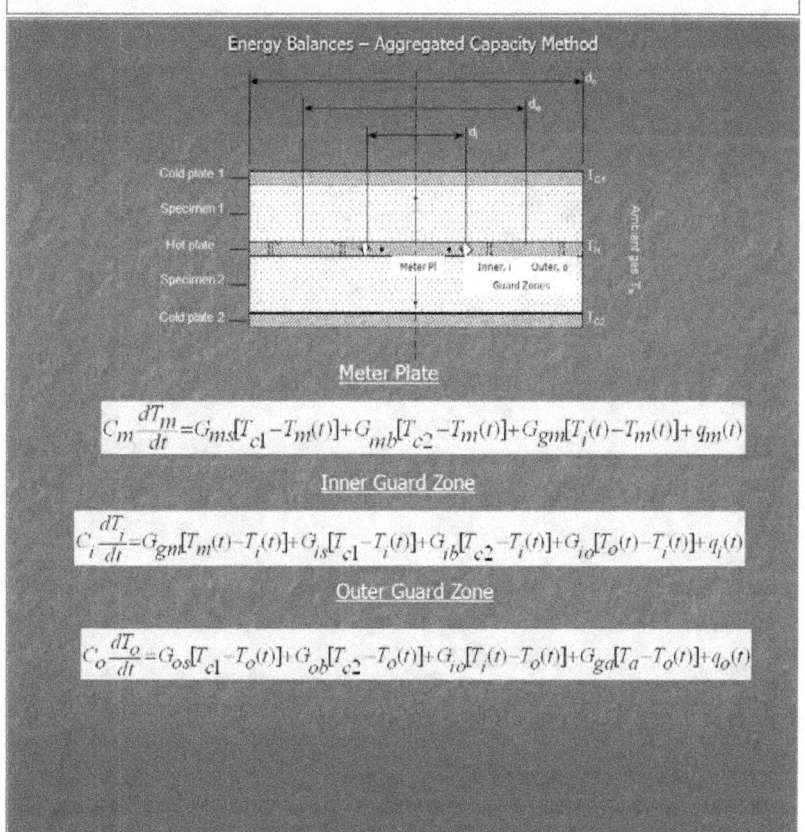

Implicit Solution Eqs

With $\dfrac{dT}{dt} \approx \dfrac{T'-T}{\Delta t}$ and rearranging, the temperature reponse Eqs are

Meter Plate

$$\left[\frac{C_m}{\Delta t} + [G_{ms} + G_{mb} + G_{gm}]\right]T'_m - G_{gm}T'_i = \frac{C_m}{\Delta t}T_m + [G_{ms}T_{c1} + G_{mb}T_{c2} + q'_m]$$

Inner Guard Zone

$$-G_{gm}T'_m + \left[\frac{C_i}{\Delta t} + [G_{is} + G_{gm} + G_{ib} + G_{io}]\right]T'_i - G_{io}T'_o = \frac{C_i}{\Delta t}T_i + [G_{is}T_{ci} + G_{ib}T_{cb} + q'_i]$$

Outer Guard Zone

$$-G_{io}T'_i + \left[\frac{C_o}{\Delta t} + [G_{os} + G_{ob} + G_{io} + G_{ga}]\right]T'_o = \frac{C_o}{\Delta t}T_o + [G_{os}T_{c1} + G_{ob}T_{c2} + G_{ga}T_a + q'_o]$$

Thermal Capacity Parameters

$$C' = \frac{mC_p}{\Delta t}$$

- ❑ Mass m by measurement
- ❑ Specific Heat C_p from Handbook
- ❑ Calculation Time Step Δt

Conductances (Thermal Bridges) between Subcomponents

$$G \approx \begin{cases} \dfrac{\lambda A_s}{L} & - \text{thru test specimen} \\ \lambda_{Al} S_\lambda & - \text{inner to outer guard} \\ h_v A_s & - \text{exposed surface} \\ \text{constant} & - \text{across gap(measured)} \end{cases}$$

Control – the "q" terms

Heating Rate:
$$q = \frac{V^2}{R_{elt}}$$

Power Supply Output:
$$V = V_{ps,\max}\left(\frac{v'}{10\ \text{vdc}}\right)$$

Controller Output:

Incremental Control Algorithm – *Finite Difference Form*

$$v' = v + K_p[e' - e] + K_i \Delta t_{sr} e' + \frac{K_d}{\Delta t_{sr}}[e' - 2e + e_{-1}]$$

$e = Error = Setpoint$ - $Process\ Variable$

$\Delta t_{sr} = Controller\ Sample\ Rate$

$K_p, K_i, \& K_d = Proportional,\ Integral,\ and\ Derivative\ Gains$

Programmed Control Algorithm

$$v' = max\left(0, min\left\{v_{hl}, \left[v + \kappa_P \Delta t_{sr} e' + \kappa_D [e' - e]\right]\right\}\right)$$

$$v_{hl} = Maximum \quad Allowable$$

$$\kappa_P, \kappa_D = Renamed \quad Gains$$

Tuning Procedure

Recall: $v' = max\left(0, min\left\{v_{hl}, \left[v + \kappa_P \Delta t_{sr} e' + \kappa_D [e' - e]\right]\right\}\right)$

- Info on Incremental Method Tuning Lacking in Literature
- Desired Response
    - "Reasonable" Initial Approach to Set Point
    - Tight control during Final Hour
- Procedure Adopted
    - Set κ_D to Zero & Manually Increase κ_P to Approach Set Point
    - Increase κ_D to Minimize R-Value Excursions in Quasi-steady Period
        - Maximum Deviation
        - RMS Deviation

Summary

- Simulation Results
  - Improved Set of Controller Gains
  - Greatly Reduced Required Time vs Traditional Operation
  - Gain Parameters Satisfactorily Used for Many Years
    - Nominal R-Values 0.4 – 4.8 C m^2/W
    - Thicknesses 13 – 230 mm
- Simulation Model Has Been Extended to NIST's 500 mm GHP

Mathematical Model for NIST's Extended-Range 500-mm GHP

- Same General Modeling Approach
- 16 PID controllers (vs 3 for 1-m GHP)
  - Not Intended for Two-Sided Operating Mode
  - Symmetry Enables Simplifying
    - 10 Controllers & Heaters
    - 44 Thermal Bridges
- Highest Uncertainty Source: Evaluation of Thermal Bridges
  - Types as for 1-m GHP
  - + SPRT, TC, TP and Heater Sheaths ("Fins")

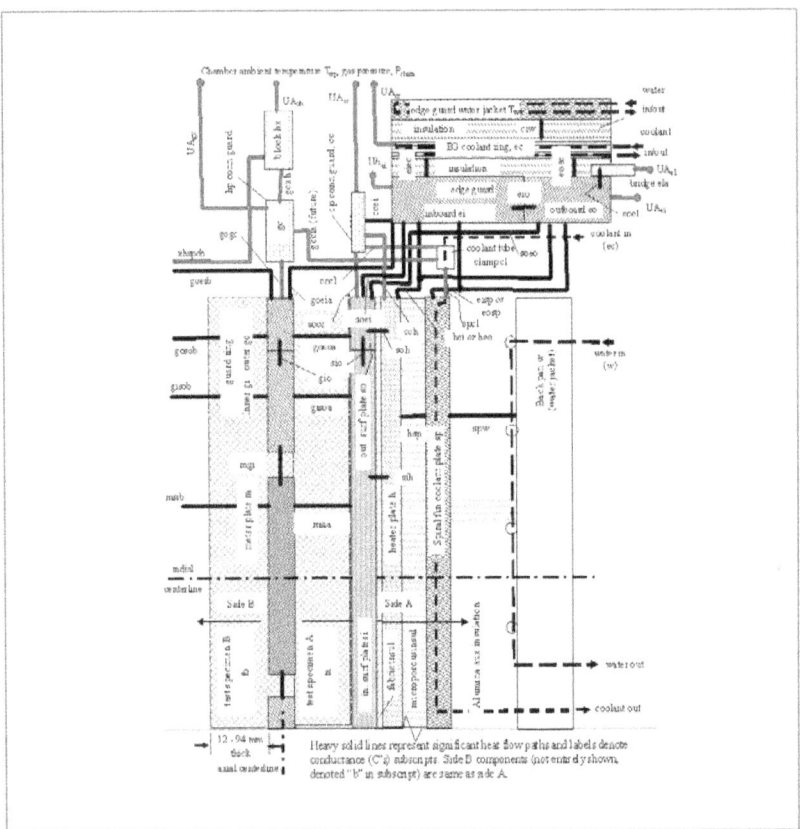

References

- W. F. Stoecker and P. A. Stoecker, *Microcomputer Control of Thermal and Mechanical Systems*, Van Nostrand Reinhold, 1988.
- W. C. Thomas and R. R. Zarr, *Thermal response simulation for tuning PID controllers in a 1016 mm guarded hot plate apparatus*, ISA Transactions 50 (2011) 504-512.

 engineering laboratory

Initial Measurement Results of the NIST 500 mm Guarded-Hot-Plate Apparatus: Automated Temperature and Pressure Control

Robert R. Zarr and William C. Thomas

2nd Operators Workshop on High-Temperature Guarded-Hot-Plate and Pipe Measurements

National Institute of Standards and Technology
Gaithersburg, Maryland, United States
March 20, 2012

NIST
National Institute of
Standards and Technology
U.S. Department of Commerce

Motivation and Goal

- Guarded-hot-plate comparisons at high-temperatures have revealed a high levels of scatter (15 %) in the data

- Few laboratory comparisons at low temperatures or at gas pressures other than atmospheric

- Provide assistance for documentary standards for guarded-hot-plate method at temperatures other than room temperature

- *Goal: Plate temperature control to within 0.01 K, or better, under steady-state conditions under different atmospheric pressures*

 engineering laboratory

Outline

- **Description of apparatus**
- **Control strategy**
  - Temperature (PID control)
  - Gas pressure
- **Results and discussion**
- **Summary**

NIST 500 mm Guarded-Hot-Plate Apparatus

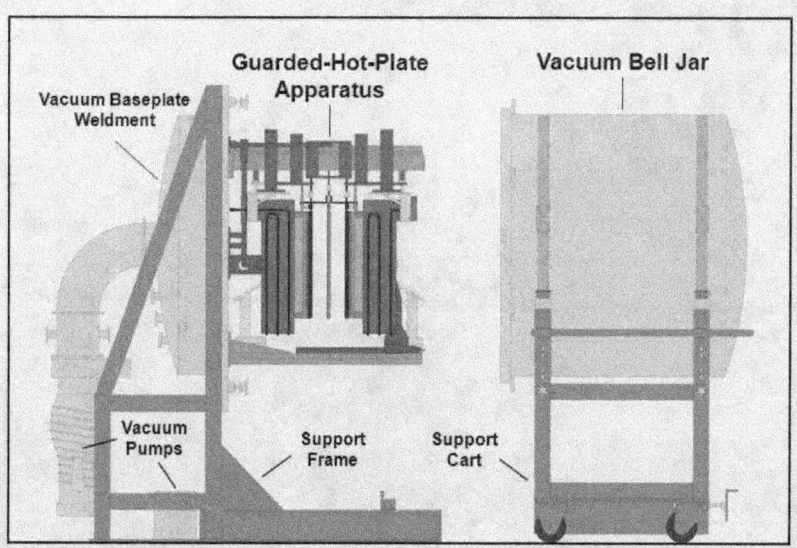

185

Apparatus and Vacuum Bell Jar

Guarded-Hot-Plate Apparatus and Vacuum Bell Jar

Guarded Hot Plate Thermometry/Heaters

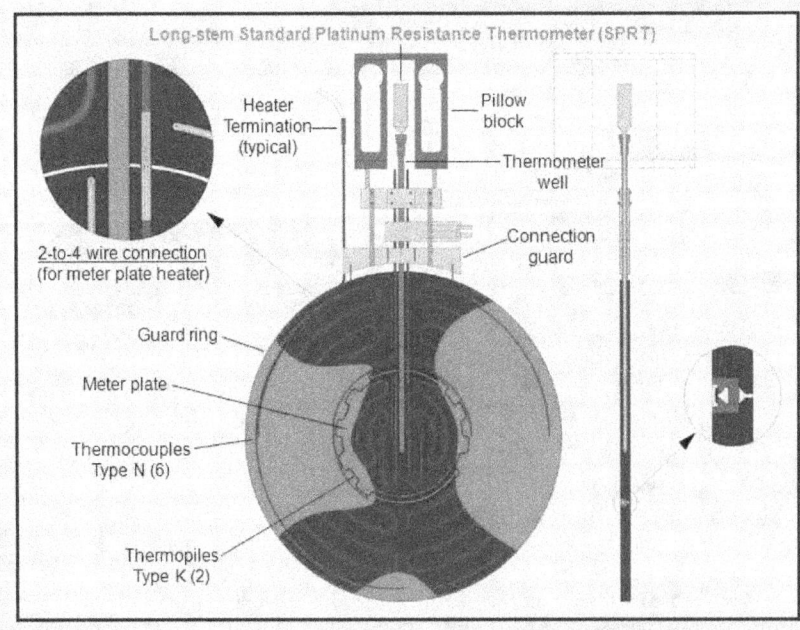

Outline

- **Motivation**
- **Description of apparatus**
- **Control strategy**
  - Temperature (PID control)
  - Gas pressure
- **Results and discussion**
- **Summary**

Temperature Control Equations

- **Discrete equivalent *PI* analog controller (*Raven, 1978*)**
  - **Positional form of the control equation**

$$m(k) = K_1 e(k) + K_2 \Delta t \left[\sum e(n)\right] + m(0)$$ (1)

  - **Incremental or velocity form of the control equation**

$$\Delta m(k) = m(k) - m(k-1) = K_1' \left[e(k) - e(k-1)\right] + K_2' \Delta t\, e(k)$$ (2)

where:
- k is current sampling interval;
- K_1', K_2' are control constants (incremental version);
- $e(k)$, $e(k-1)$ are errors from setpoint for current and previous intervals; and,
- Δt is the sampling interval (s)

- **16 separate heaters ➔ 16 PI control equations (32 control constants *(K_1', K_2')* to be determined)**

Gas Pressure Control - Schematic

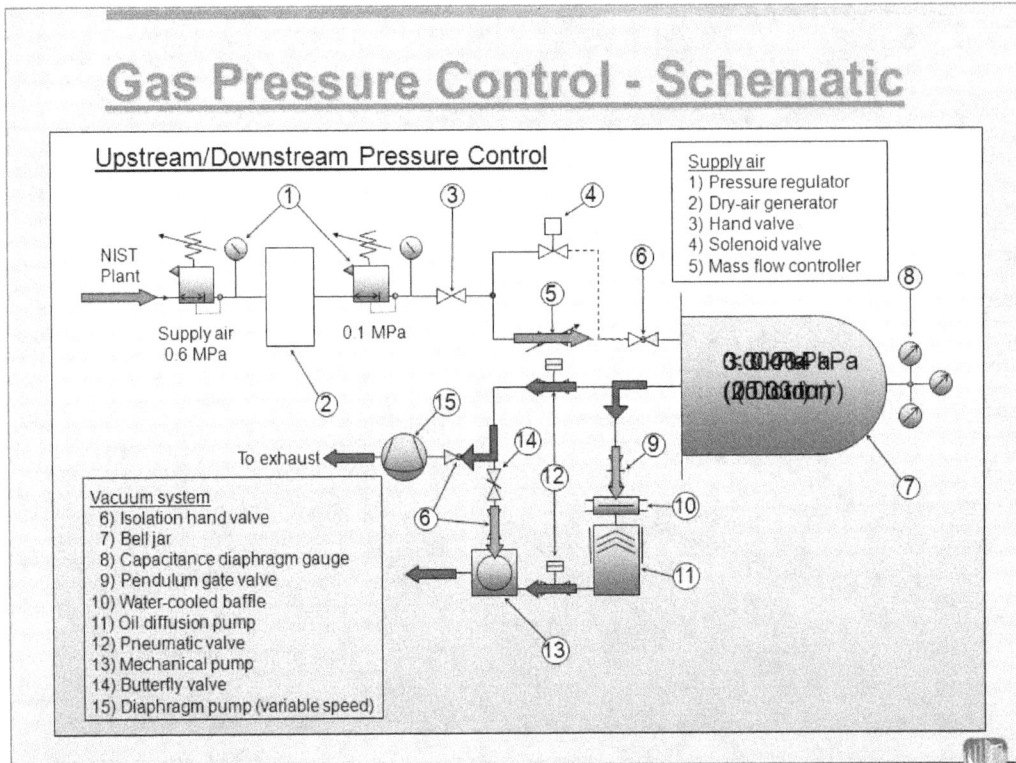

Upstream/Downstream Pressure Control

Supply air
1) Pressure regulator
2) Dry-air generator
3) Hand valve
4) Solenoid valve
5) Mass flow controller

NIST Plant

Supply air 0.6 MPa

0.1 MPa

To exhaust

Vacuum system
6) Isolation hand valve
7) Bell jar
8) Capacitance diaphragm gauge
9) Pendulum gate valve
10) Water-cooled baffle
11) Oil diffusion pump
12) Pneumatic valve
13) Mechanical pump
14) Butterfly valve
15) Diaphragm pump (variable speed)

Data Acquisition Schematic

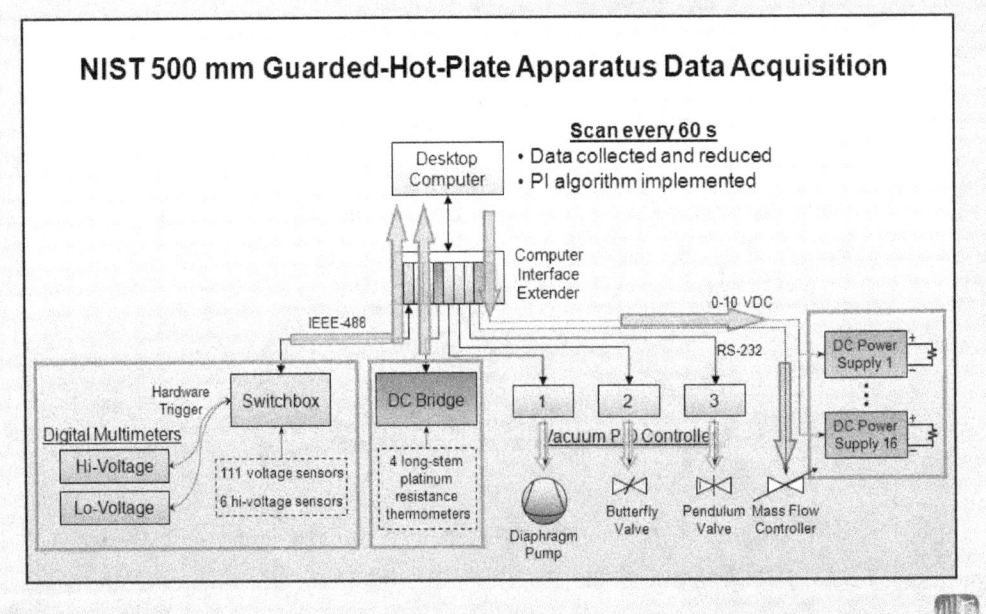

NIST 500 mm Guarded-Hot-Plate Apparatus Data Acquisition

Desktop Computer

Scan every 60 s
- Data collected and reduced
- PI algorithm implemented

Computer Interface Extender

0-10 VDC

IEEE-488

RS-232

DC Power Supply 1

DC Power Supply 16

Hardware Trigger

Switchbox

DC Bridge

1 2 3

Vacuum Pump Controller

Digital Multimeters

Hi-Voltage

Lo-Voltage

111 voltage sensors
6 hi-voltage sensors

4 long-stem platinum resistance thermometers

Diaphragm Pump

Butterfly Valve

Pendulum Valve

Mass Flow Controller

Outline

- **Motivation**
- **Description of apparatus**
- **Control strategy**
  - Temperature (PID control)
  - Gas pressure
- **Results and discussion**
- **Summary**

Meter Plate Temperature Control
Chamber at atmospheric pressure (uncontrolled)

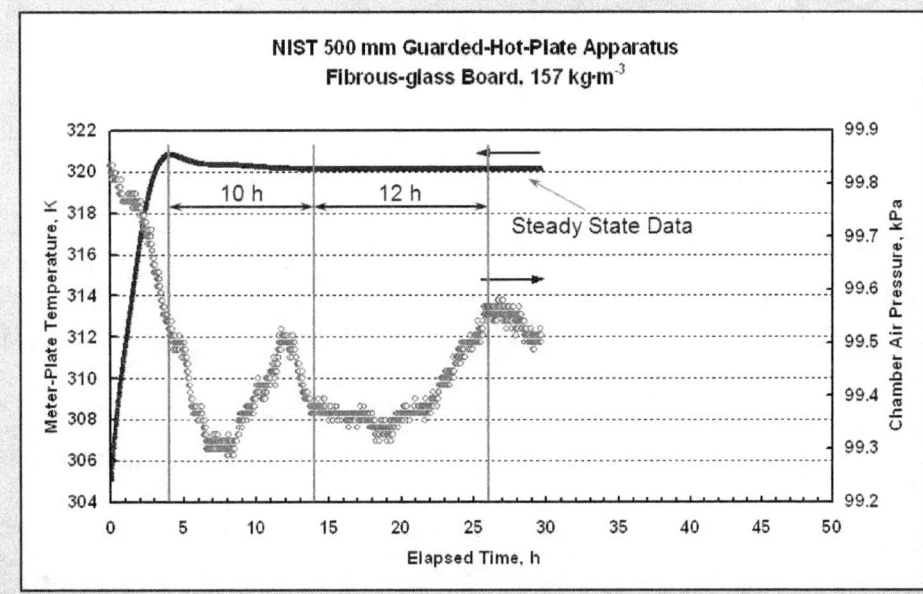

NIST 500 mm Guarded-Hot-Plate Apparatus
Fibrous-glass Board, 157 kg·m^{-3}

189

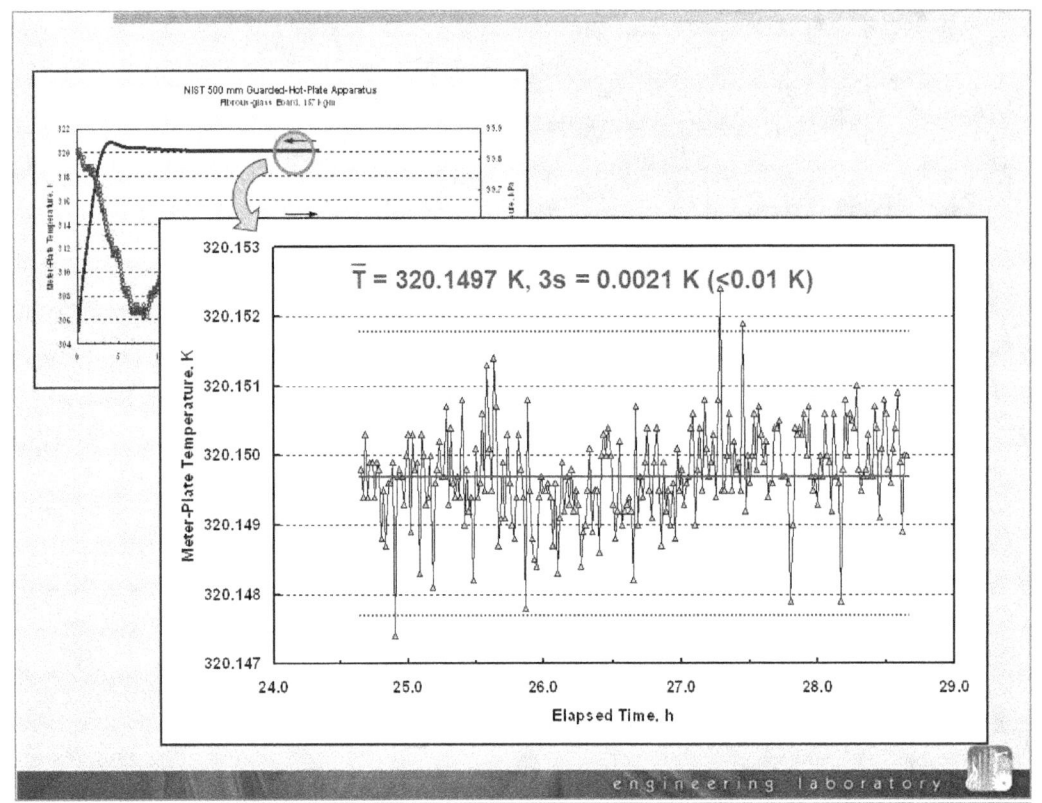

$\overline{T}$ = 320.1497 K, 3s = 0.0021 K (≤0.01 K)

Meter Plate Temperature Control
Chamber under reduced pressure (controlled)

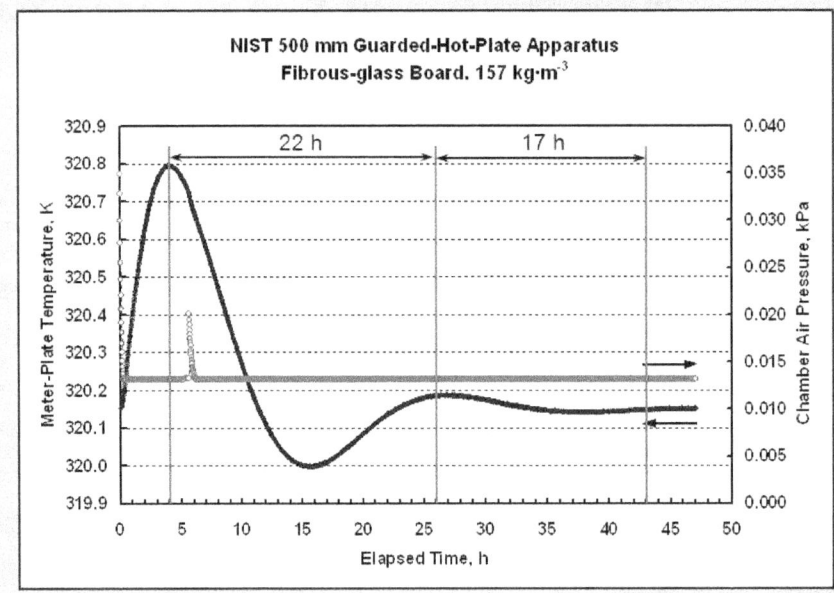

NIST 500 mm Guarded-Hot-Plate Apparatus
Fibrous-glass Board, 157 kg·m⁻³

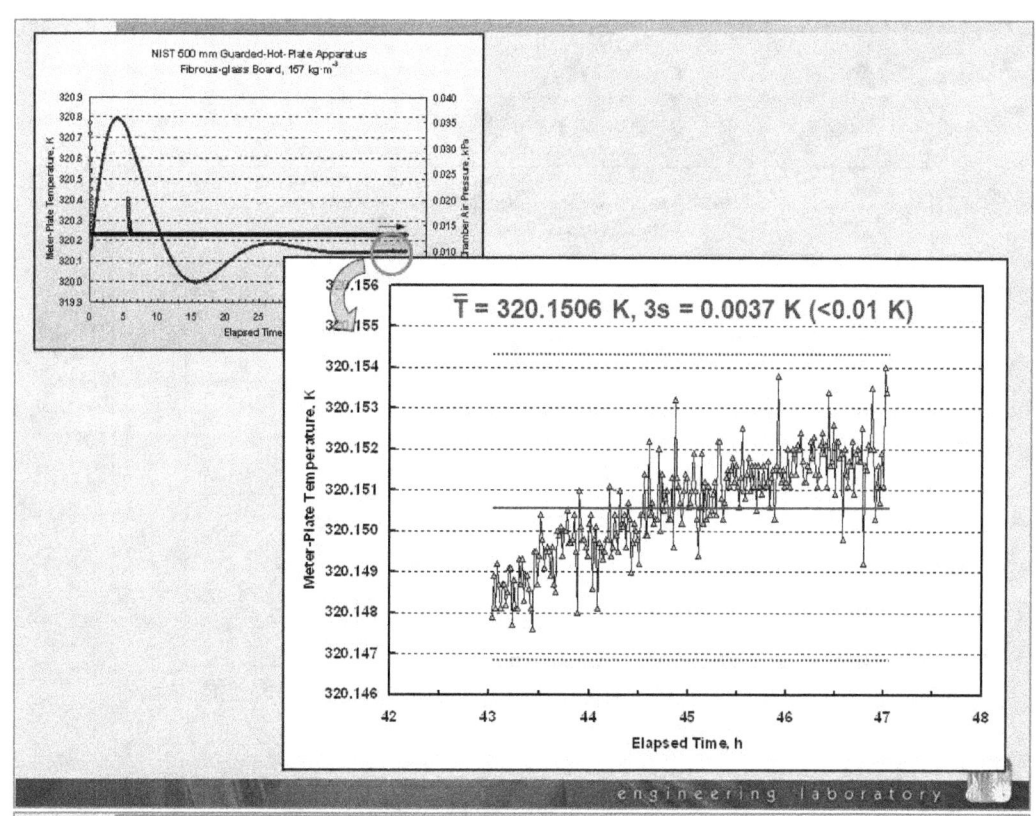

$\overline{T} = 320.1506$ K, 3s = 0.0037 K (<0.01 K)

Thermal Conductivity Data
λ as a function of T_m ($\Delta T = 20$ K)

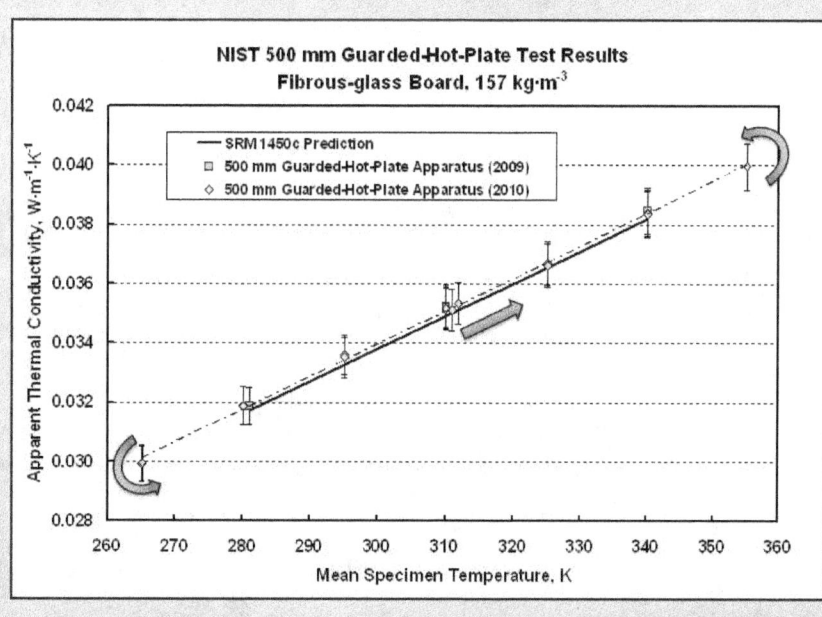

Thermal Conductivity Data
λ as a function of chamber air pressure (↓)

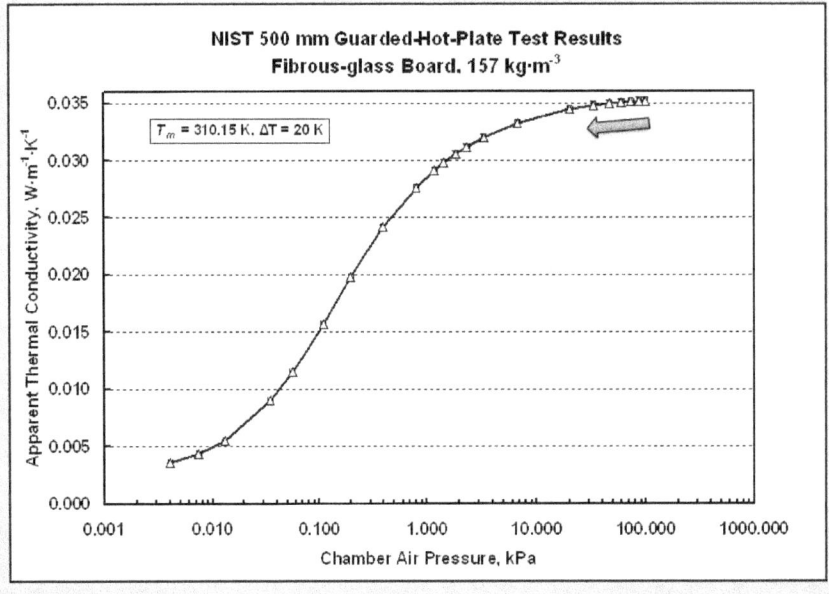

NIST 500 mm Guarded-Hot-Plate Test Results
Fibrous-glass Board, 157 kg·m⁻³

T_m = 310.15 K, ΔT = 20 K

Discussion

- Control stability for meter plate temperature requires at least 20 h

- Steady-state data (4 h):
    - Average value is within 0.002 K, or less, of setpoint
    - Range < 0.006 K (less than goal of 0.01 K)

- Pressure control < 0.001 torr

- Thermal conductivity
    - Temperature data within 1 % of SRM 1450c prediction (from NIST 1016 mm guarded-hot-plate apparatus)
    - Pressure data consistent with previous NIST-Boulder data on SRM 1450b

Summary and Future Work

- Control stability achieved within stated goal
- Additional work may be required for the control constants at low pressures
- Thermal conductivity results consistent with previous SRM 1450 results (temperature and pressure)

- *Future work: Inter-lab comparisons with other NMIs (National Metrology Institutes) at :*
  - *Elevated temperatures*
  - *Low gas pressures*

engineering laboratory

2nd Operators Workshop on High-Temperature Guarded-Hot-Plate and Pipe Measurements

High-Temperature Thermal Insulation
Industry Needs

Thomas Whitaker
Chairman, ASTM Committee C16 on Thermal
Insulation

Control System Considerations
Steady State Issues

- PID Control
  - PID model for the NIST high temperature plate
  - Application of that model to the NIST high temperature plate
  - Pipe PID control
    - IIG R&D Center Pipe
    - 3 temperature controllers with thermocouple input
    - Output of the controller connected to the main power supply is a proportional DC voltage.

Power Control

- Use PID control to get the pipe to the target temperature, then hold
- After a minimum 2 hour period, switch the controller to "Constant Power" mode. Set the output equal to the average output from the previous 2 hours.
  - DC volts
  - AC volts

Typical Test

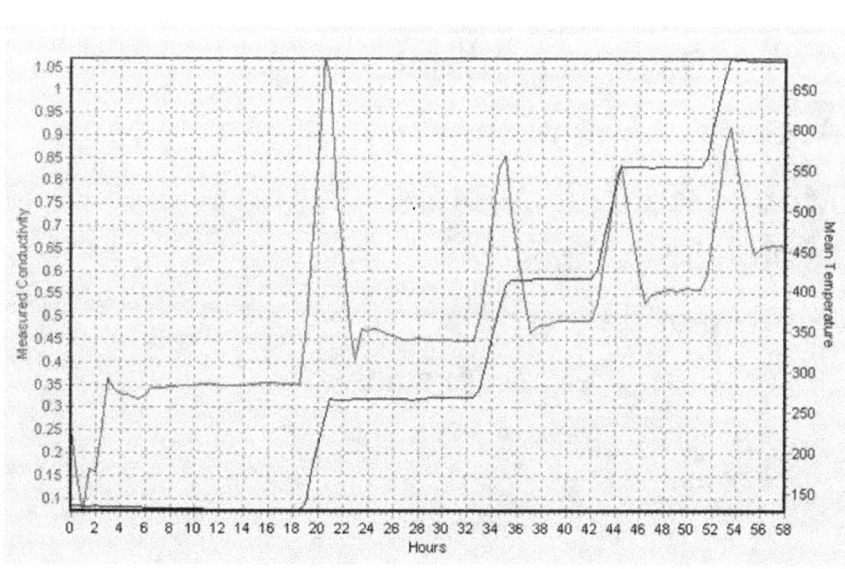

- Setting the PID values was not trivial
  - The controller Auto Tune worked well at the lower temperatures but not well at the higher temperatures
- I use different PID settings for each target temperature.

Steady State

- Collect Data every minute
- Average at 5 minutes, then at 30 minutes
- Monitor for changes in
  - Hot surface temp
  - Guard offset
  - Measured thermal conductivity
    - Change <0.5% over a 2 hour period
    - Monotonically decreasing or increasing

Questions & Discussion

PID temperature control versus locking power (temperature drifts)

Level of control at different temperatures (how precise?)

How many data points per run

How to define steady-state

Estimating Measurement Uncertainty of Thermophysical Properties

Robert Zarr

National Institute of Standards and Technology
Engineering Laboratory
Energy and Environment Division

2nd Operators Workshop on High-Temperature
Guarded-Hot-Plate and Pipe Measurements

Gaithersburg, Maryland, United States
March 20, 2012

ASTM INTERNATIONAL
Standards Worldwide

NIST
National Institute of
Standards and Technology
U.S. Department of Commerce

"Classic" NBS Measurement Uncertainty Statement

Cyril H. Meyers (1930s)

H. H. Ku

- 1972 – Dr. H. H. Ku, NBS Statistician, reported that C. H. Meyers, NBS physicist, originally stated the uncertainty for his measurements on the heat capacity of ammonia as follows:

  "We think our reported value is good to 1 part in 10,000: we are willing to bet our own money at even odds that it is correct to 2 parts in 10,000. Furthermore, if by any chance our value is shown to be in error by more than 1 part in 1000, we are prepared to eat the apparatus and drink the ammonia."

- The statement was not approved by the NBS Editorial Board and is only preserved anecdotally (by Ku).

Complications – Multitude of...

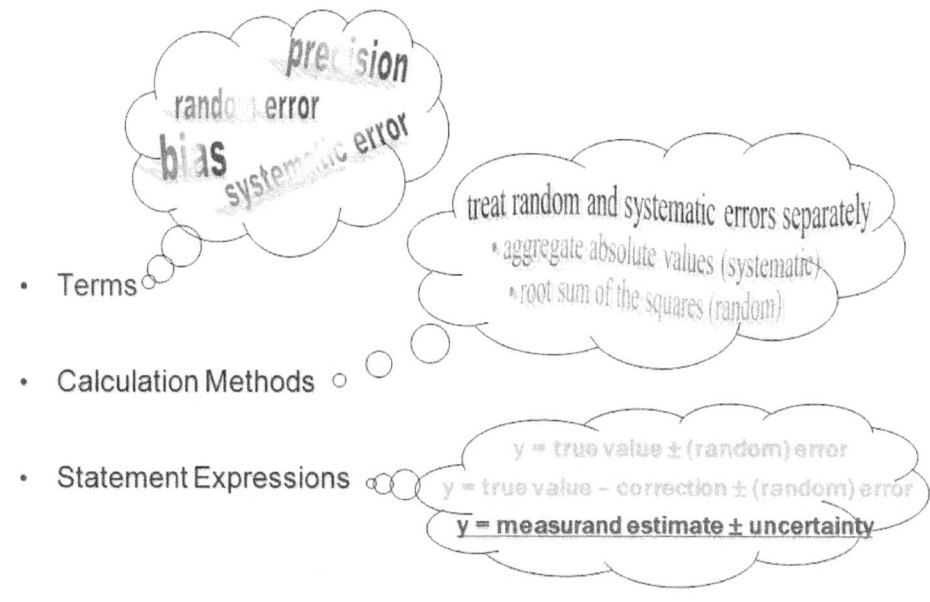

- Terms

- Calculation Methods

- Statement Expressions

Timeline
Standardization of Measurement Uncertainty

- 1978 – Comité International des Poids et Measures (CIPM)
  formally requested Bureau International des Poids et
  Measures (BIPM) to address the problem in conjunction
  with the national standards laboratories

- 1980 – Recommendation INC-1 (BIPM Working Group)

  Transferred to ➔ ISO Technical Advisory Group on Metrology
  (TAG 4), Working Group 3 (BIPM, IEC, ISO, and OIML)

- 1993 – ISO Publication – "*Guide to the Expression of Uncertainty
  of Measurement*" (**GUM**)

- 1994 – NIST Technical Note 1297 by B.N. Taylor and C.E. Kuyatt
  (NIST Policy for all publications)
  http://physics.nist.gov/Pubs/guidelines/TN1297/tn1297s.pdf

- 2008 – BIPM CGM 100:2008 (GUM 1995 with minor corrections)

GUM Philosophy and Approach

Philosophy
- *International consensus* – uniform method so that measurements in different countries can be compared (analogous to *SI* units)
- *Universal* – applicable to all kinds of quantitative measurements and to all types of input data

Approach
- GUM does not distinguish between random and systematic components
- Instead, evaluations are classified as either:
    - Type A – those evaluations by statistical analysis of the data, or
    - Type B – those evaluations by other means.

Type A Evaluation of Standard Uncertainty

- Repeated independent observations
    - Sample mean

$$x_i = \bar{X}_i = \frac{1}{n}\sum_{k=1}^{n} X_{i,k}$$

Mean "best available estimate"

The normal distribution

- Standard uncertainty is the <u>standard deviation of the mean</u>

$$u(x_i)_A = s(\bar{X}_i) = \frac{s_x}{\sqrt{n}}$$

where s_x is the sample standard deviation

- Linear regression (method of least squares) } not adequately discussed in GUM
- Analysis of variance (ANOVA)   } "you are on your own"

Type B Evaluation of Standard Uncertainty

- Usually based on scientific judgment, which may include:
  - Calibration data from another laboratory →
  - Previous measurement data
  - Manufacturer specification
  - Experience with, or general knowledge of, the behavior and property of relevant materials and instruments

- Reasonable default model – estimate only bounds (upper and lower limits) by assuming uniform distribution (state your assumption)

- Standard uncertainty

$$u_B = \frac{a}{\sqrt{3}}$$

where

$$a = (a_+ - a_-)/2$$

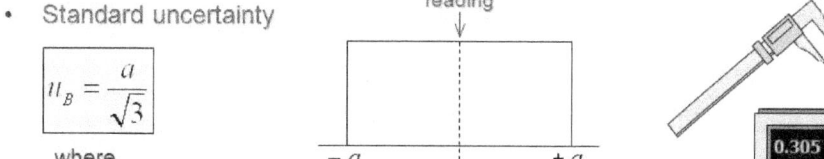

The uniform (or rectangular) distribution

Discussion – Type A Versus Type B

- Which type of evaluation is better? A or B or both? (it depends on your experiment)

- Type A
  - At lower levels of calibration chain, where reference standards assumed known, the uncertainty may be a single Type A standard uncertainty evaluated from the pooled experimental data (GUM)
  - Complicated measurement situation consider obtaining guidance from a statistician (NIST Technical Note 1297)

- Type B
  - It should be recognized that a Type B evaluation can be as reliable as a Type A, especially when the evaluation is based on a small number of observations (GUM)
  - Type B generally provides metrology traceability (calibration certificate)

- Note: There is not always an obvious correspondence between A and B evaluations and the classical classification of "random" and "systematic"

    On the other hand, "don't throw away old concepts, keep both."

Combined Standard Uncertainty, $u_c(y)$

- Measurand, Y, and input quantities, X_i $$Y = f(X_1, X_2, ..., X_N)$$

- Output estimate, y, and input estimates, x_i $$y = f(x_1, x_2, ..., x_N)$$ "numbers"

- Law of propagation of uncertainty (assuming no correlation between x_i values)

$$u_c(y) = \sqrt{c_{x_1}^2 u^2(x_1) + c_{x_2}^2 u^2(x_2) + ... + c_{x_N}^2 u^2(x_N)}$$

with

$$c_{x_i} = \frac{\partial f}{\partial x_i} = sensitivity\ coefficient$$

Good news: combine individual uncertainties whether arising from Type A or Type B evaluations

---

Expanded Uncertainty, U

- After estimating the combined standard uncertainty ($u_c(y)$) of a measurement result, the final task is to compute the expanded uncertainty, denoted as U
- The expanded uncertainty is computed using the formula:

$$U = k u_c$$

- The coverage factor typically denoted, k, and obtained from the Student's t distribution, controls the probability with which the measurement result ± its expanded uncertainty will contain the measurand
- At NIST, k is typically taken to be 2 (NIST Technical Note 1297)

Note: not everyone at NIST uses a value of 2 for k.

Summary of Steps

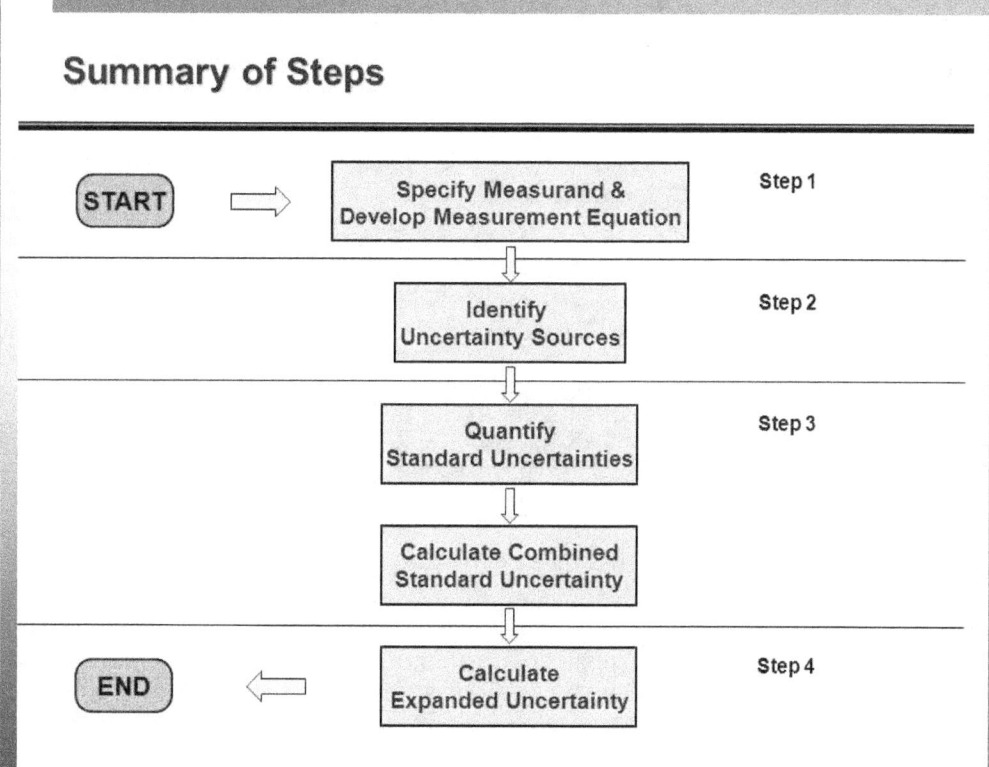

START ⟹ **Specify Measurand & Develop Measurement Equation** — Step 1

Identify Uncertainty Sources — Step 2

Quantify Standard Uncertainties — Step 3

Calculate Combined Standard Uncertainty

END ⟸ **Calculate Expanded Uncertainty** — Step 4

Example – Thermal Conductivity

1a) Specify measurand (particular quantity subject to measurement)

Example – (apparent) thermal conductivity of thermal insulation in air at 24 °C, 25 K temperature difference, 50 % relative humidity, 1 kPa clamping force, 101.32 kPa gas pressure

1b) Define measurement equation (estimate for measurand)

Fourier heat conduction equation (1-D, algebraic form)

$$Q = \lambda A \frac{\Delta T}{L} \implies \lambda = \frac{QL}{A \Delta T}$$

where input quantities Q, L, A, and ΔT are the specimen heat flow [W]; specimen thickness (plate separation) [m]; meter area [m^2]; and, temperature difference [K], respectively

Example – Thermal Conductivity (continued)

2) Identify uncertainty sources (cause and effect chart)

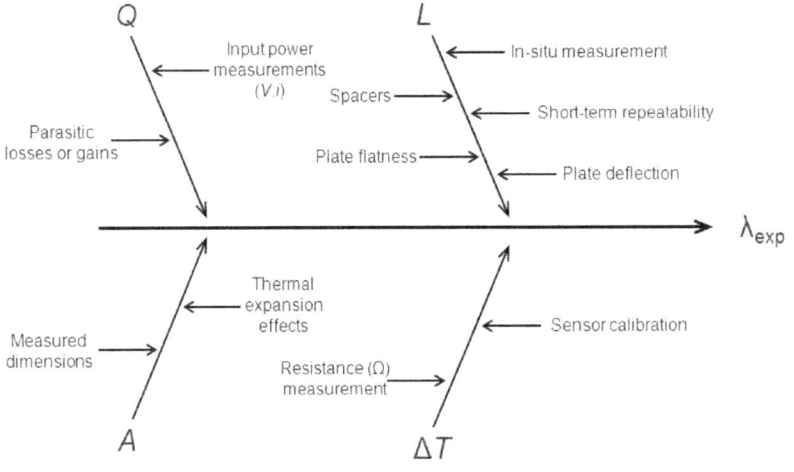

NIST 1016 mm guarded-hot-plate apparatus

Example – Thermal Conductivity (continued)

3a) Quantify standard uncertainties

"Uncertainty Budget"

NIST 1016 mm guarded-hot-plate apparatus (single-sided mode)

fibrous-glass insulation 25 4 mm specimen

TABLE III. SUMMARY OF STANDARD UNCERTAINTY COMPONENTS

Source of Uncertainty	Value	Evaluation
Meter area (A)	2.5×10^{-2} m	B
Plate dimensions	0.025 mm	B
Thermal expansion coefficient	2.4×10^{-6} K^{-1}	B
Temperature measurement	0.086 K	B
Thickness (L)	38×10^{-6} m	B
In-situ linear position measurement system	20×10^{-6} m	B
Multiple observations	19×10^{-6} m	A
System uncertainty	5×10^{-6} m	B
Dimensions of fused-quartz spacers	1.9×10^{-6} m	B
Repeated observations	1.1×10^{-6} m	A
Caliper uncertainty	1.5×10^{-6} m	B
Short-term repeatability	6.4×10^{-6} m	A
Plate flatness	7.9×10^{-6} m	B
Repeated observations	2.3×10^{-6} m	A
Coordinate measuring machine (CMM) uncertainty	5.1×10^{-6} m	B
Plate deflection under axial loading of cold plate	31×10^{-6} m	B
Temperature difference (ΔT)	0.086 K	B
Measurement (T, T)	0.058 K	B
Digital multimeter (DMM) uncertainty	0.058 K	B
Platinum resistance thermometer (PRT) regression analyses for calibration data	0.005 K	A
Calibration of PRTs	0.005 K	B
Miscellaneous sources	0.019 K	B
Contact resistance	0.017 K	B
Sampling of planar plate temperature	0.015 K	B
Axial temperature variations	0.011 K	B
Heat flow (Q)	0.009 W	B
Repeated observations	0.0006 W	A
Direct current power measurement (Q)	0.0017 W	B
Standard resistor calibration	2.5×10^{-3} Ω	B
Standard resistor drift	(in progress)	---
PRT power input	0.0001 W	B
Voltage measurement – standard resistor	15 μV	B
Voltage measurement – heater	3 mV	B
Parasitic heat flows (ΔQ)	0.087 W	B
Guard-gap (Q)	(Reference [13])	
Auxiliary insulation (Q)	(Reference [13])	
Edge effects (Q)	(Reference [13])	

Dominant components of uncertainty

Example – Thermal Conductivity (continued)

3b) Compute combined standard uncertainty, $u_c(\lambda)$

$$u_c(\lambda) = \sqrt{\left[\frac{L}{A\,\Delta T}\right]^2 u_Q^2(Q) + \left(\frac{Q}{A\,\Delta T}\right)^2 u_L^2(L) + \left(\frac{-QL}{A^2\,\Delta T}\right)^2 u_A^2(A) + \left(\frac{-QL}{A\,\Delta T^2}\right)^2 u_{\Delta T}^2(\Delta T)}$$

with

$$c_Q = \frac{\partial \lambda}{\partial Q}, \quad c_L = \frac{\partial \lambda}{\partial L}, \quad c_A = \frac{\partial \lambda}{\partial A}, \; and, \; c_{\Delta T} = \frac{\partial \lambda}{\partial(\Delta T)}$$

Relative combined standard uncertainty, $u_{c,r}(\lambda)$

$$u_{c,r}(\lambda) = \frac{u_c(\lambda)}{\lambda} = \sqrt{\left(\frac{u_Q(Q)}{Q}\right)^2 + \left(\frac{u_L(L)}{L}\right)^2 + \left(\frac{u_A(A)}{A}\right)^2 + \left(\frac{u_{\Delta T}(\Delta T)}{\Delta T}\right)^2}$$

Example – Thermal Conductivity (continued)

4) Compute expanded uncertainty, U, for statement

Example

$\lambda = (0.04500 \pm 0.00041)$ W·m^{-1}·K^{-1}, where the number following the symbol ± is the numerical value of an expanded uncertainty $U = ku_c$, with U determined from a combined standard uncertainty (i.e., estimated standard deviation) $u_c = 0.00020$ W·m^{-1}·K^{-1} and a coverage factor $k = 2$. Because it can be assumed that the possible estimated values of the standard are approximately normally distributed with approximate standard deviation u_c, the unknown value of the standard is believed to lie in the interval defined by U with a level of confidence of approximately 95 %.

Relative expanded uncertainty, U_r

$U_r = (0.00041/0.04500) \times 100 = 0.9\%$ (NIST guarded-hot-plate measurements for customers are typically rounded up to nearest 0.5%)

Summary

- International consensus – analogous to *SI* units
- Uniform approach to the expression of measurement uncertainty
- Now in effect at:
  - National metrology institutes
  - Metrology laboratories
  - Accredited laboratories
  - Other laboratories – for example, nuclear, forensics, etc.

- ISO Guide 98-3, "Uncertainty of measurement – Part 3: Guide to the expression of uncertainty in measurement (GUM:1995), 2008
- ISO Guide 99, "International vocabulary of metrology – Basic and general concepts and associated terms (VIM), 2007

Both documents now available **free** for download at:

http://www.bipm.org/utils/common/documents/jcgm/JCGM_100_2008_E.pdf

http://www.bipm.org/utils/common/documents/jcgm/JCGM_200_2012.pdf

Uncertainty Analysis using the GUM and GUM Supplement

Blaza Toman

Statistical Engineering Division

ITL, NIST

1

Outline

- Example – Measuring Thermal Conductivity
- Uncertainty - definitions
- Probability – short review
- Uncertainty Analysis using GUM Supplement
- Uncertainty Analysis using GUM
- Modeling trends and uncertainty based on repeated measurements
- Conclusions

2

Measuring Thermal Conductivity
using the Guarded-Hot-Plate apparatus

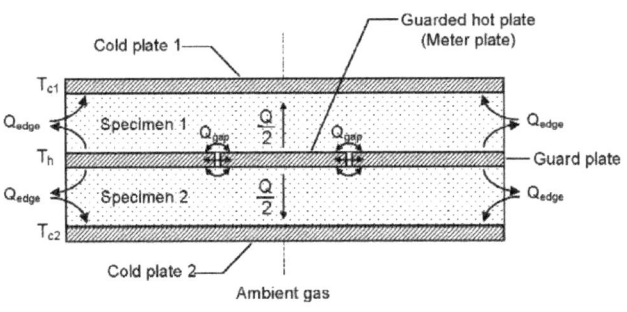

Measuring Thermal Conductivity

Equation $$\lambda = \frac{\frac{Q}{2} \, L_{Average}}{A \, \Delta T_{Average}}$$

- Q is the time rate of one-dimensional heat flow through the meter area of both specimens

- A is the meter area

- $L_{Average}$ is the thickness

- $\Delta T_{Average} = T_h - T_c$ is the temperature difference

Other Considerations

From previous work it is known that thermal conductivity λ may be a function of

1. bulk density ρ - not part of the measurement equation

2. Temperature $\quad T_m = \dfrac{T_h + T_c}{2}$

As in equation $\quad \lambda = a_0 + a_1\rho + a_2 T_m$

Example – NIST SRM 1450d
Fibrous-Glass Board
Thermal conductivity from 280K to 340K [1]

Sample data from this experiment

T_m (K)	ρ (kg m^{-3})	T_h (K)	T_{c1} (K)	T_{c2} (K)	Q/2 (W)	A (m^2)	L (mm)	λ (Wm^{-1}K^{-1})
280	113.5	292.50	267.50	267.50	3.895	0.12980	25.93	0.03112
295	114.3	307.50	282.50	282.50	4.050	0.12989	26.02	0.03245
295	119.0	307.50	282.50	282.50	4.086	0.12989	25.82	0.03249
340	123.8	352.50	327.49	327.50	4.736	0.13016	25.85	0.03782

The total data set included 15 measurements in a 3x5 factorial experiment with three density ρ (low, mid, high) levels and five temperature T_m levels (280, 295, 310, 325, 340).

Thermal conductivity calculation

For each experimental observation (row in the table) λ (called the measurand) is computed according to

$$\lambda = \frac{\frac{Q}{2} L_{Average}}{A \, \Delta T_{Average}}$$

This gives the value, we need an accompanying uncertainty

Uncertainty according to the GUM [2]

- measurement uncertainty "reflects the lack of exact knowledge of the value of the measurand".

- The corresponding state of knowledge is best described by means of a probability distribution over the set of possible values for the measurand.

- Standard uncertainty is the standard deviation of this distribution, expanded uncertainty is the 95% probability interval.

So what exactly is a probability distribution?

Imagine flipping a coin 2 times and recording Y=number of heads in two flips

Y = 0 with probability 0.25

Y = 1 with probability 0.5

Y = 2 with probability 0.25

This is an example of a probability distribution

Another example – Human Height

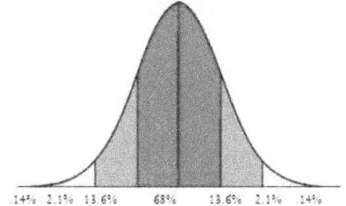

Usually described using a Gaussian probability distribution (bell curve)

For American males - mean is 70 in, standard deviation is 3 in.

Probability that randomly selected male's height is in (63in, 73in) is 0.68

What exactly is uncertainty then?

Without considering ρ and T_m for the moment,

Recall that

$$\lambda = \frac{Q/2 \; L_{Average}}{A \; \Delta T_{Average}}$$

We need to obtain a <u>probability distribution</u> for λ, or at least the <u>standard deviation</u> of this distribution.

But isn't λ a constant? Not if the input quantities have uncertainties.

Uncertainty using GUM Supplement [3]

- For each input variable (Q, L, A, ΔT) measured with uncertainty obtain a probability distribution using data and/or uncertainty budget.

- Generate values of (Q, L, A, ΔT) from the distributions and compute λ

- This gives draws from the distribution of λ

  Compute mean and standard distribution to get estimate of λ and its standard uncertainty.

Distributions of input quantities
SRM 1450d example

Gaussian with

1. mean based on measurements in Table 13 (p. 55)
2. standard deviation obtained via detailed
   uncertainty analysis given in Annex 4.

For example, for T_m = 340K, ρ = 123.8 kg·m^{-3}

$$\frac{Q}{2} \sim Gaussian(mean = 4.76, sd = 0.0074)$$

All Distributions
for T_m = 340K, ρ = 123.8 kg·m^{-3}

$$\frac{Q}{2} \sim Gaussian(mean = 4.76, sd = 0.0074)$$

$$L \sim Gaussian(mean = 0.02585, sd = 0.000065)$$

$$A \sim Gaussian(mean = 0.13016, sd = 0.000043)$$

$$\Delta T \sim Gaussian(mean = 25.005, sd = 0.077)$$

Code for GUM Supplement analysis

```
m<-10000
tavmean<-352.5 – 327.495
tavsd<-0.077
lmean<-0.02585
lsd<-0.000065
amean<-0.13016
asd<-0.000043
qmean<-4.76
qsd<-0.0074
tav<-rnorm(m,mean=tavmean,sd=tavsd)
l<-rnorm(m,mean=lmean,sd=lsd)
a<-rnorm(m,mean=amean,sd=asd)
q<-rnorm(m,mean=qmean,sd=qsd)
lamda<-q*l/a/tav
plot(density(lamda))
mean(lamda)
sd(lamda)
```

Code in R language: R Development Core Team.
R: A language and environment for statistical computing.
R Foundation for Statistical Computing,
Vienna 2003, ISBN 3-900051-00-3

15

GUM Supplement Results for λ

T_m = 340K, ρ = 123.8 kg·m^{-3}

- The mean = 0.03781 W·m^{-1}·K^{-1}

- The standard deviation (standard uncertainty) = 0.0001615

- The relative standard uncertainty $\dfrac{0.0001615}{0.03781} = 0.0043$

16

214

Uncertainty Analysis via GUM

 GUM method <u>approximates</u> the standard deviation of the probability distribution of the measurand, using a Taylor series expansion which linearizes the measurement equation.

 This results in a formula for the standard deviation (uncertainty):

$$sd(\lambda) = \sqrt{c_Q^2 sd^2(Q) + c_{\Delta T}^2 sd^2(\Delta T) + c_L^2 sd^2(L) + c_A^2 sd^2(A)}$$

where the coefficients are partial derivatives such as

$$c_Q = \frac{\partial \lambda}{\partial Q} = \frac{L}{A \Delta T}\bigg|_{L=25.85,\, A=0.13016,\, \Delta T=25.001}$$

17

Result via GUM for λ
for T_m = 340K, ρ = 123.8 kg·m⁻³

 The mean = 0.03782 W·m⁻¹·K⁻¹
 The standard deviation (standard uncertainty) = 0.0001626
 The relative standard uncertainty = 0.0043

Recall results via GUM Supplement:

    mean = 0.03781 W·m⁻¹·K⁻¹

    relative standard uncertainty = 0.0043

18

Additional statistical modeling

- Density ρ and temperature T_m are not directly part of the measurement equation, but are known to have an effect on λ. This relationship is evaluated via the 3x5 experiment.

- The "data" - at each experimental setting – are the GUM results for λ

For example at $T_m = 340K$:
  low ρ : mean = 0.03753, sd = 0.1607 (0.43%)
  med ρ : mean = 0.03761, sd = 0.1606
  high ρ : mean = 0.03782, sd = 0.1626

19

Relationships - ρ

- Based on analysis done in NIST SP260-173 - there is <u>no</u> <u>systematic</u> relationship between ρ and λ. But there is random variability in λ due to ρ. This can be accounted for in the uncertainty.

- For example at $T_m = 340K$, there are three Gaussian curves. These can be combined [4] into a single one which has mean $= 0.03765$, std $= 0.1998$ (0.53%).

20

Relationships - T_m

 Based on analysis done in NIST SP260-173 - there is a <u>linear relationship</u> between T_m and λ, through the origin.

$$\lambda = a_1 T_m$$

 Need to estimate the coefficient a_1

Estimation of a_1

At each temperature T_m, we have a value of λ,
an uncertainty and a probability distribution (GUM Supplement)

Monte Carlo Procedure:

 At each of the $5\,T_m$ values, generate a value of λ, from the corresponding Gaussian.

For example for T_m = 340K: Gaussian(mean=0.03765, sd=0.1998).

 Fit the linear regression

 Repeat.

R program

```
x = c(280, 295, 310, 325, 340)
y.mu = c(0.030997, 0.0325203, 0.0342013, 0.0358987, 0.0376477)
y.sigma =c(0.0001658, 0.0001599, 0.0001677, 0.0001957, 0.0001998)
n = length(y.mu)
m = 1000
ab = array(dim=c(m,1))
ac = array(dim=c(m,5))
ad = array(dim=c(m,15))
for (i in 1:m)
{ y = rnorm(n, mean=y.mu, sd=y.sigma)
  z = lm(y~x-1)
  ab[i,] = coefficients(z)
  ad[i,] = predict(z,interval = "prediction",level =0.95)}
  plot(density(ab[,1]))
  mean(ab[,1])
  sd(ab[,1])
```

Results

$$\lambda = \left(1.10497 \times 10^{-4}\right) T_m$$

with uncertainty of the slope (2.467×10^{-7})

Predicted value at $T_m = 340$ K from this model is $0.038 \ W{\cdot}m^{-1}{\cdot}K^{-1}$, with uncertainty 0.00023 (0.60%)

This uncertainty includes components from Q, L, A, ΔT, ρ, and the regression on T_m

Conclusion

- Demonstrated both traditional GUM analysis and GUM Supplement analysis on example from SRM 1450d.

- GUM Supplement analysis of uncertainty is straightforward and enables more complex statistical modeling which can produce estimates of systematic effects such as linear trends, as well as account for additional uncertainty.

25

References

- [1] NIST SP 260-173
- [2] JCGM 100:2008. Evaluation of measurement data – guide to the expression of uncertainty in measurement. Joint Committee for Guides in Metrology (JCGM), Sèvres, France (2008) http://www.bipm.org/en/publications/guides/gum.html.
- [3] JCGM 101:2008. Evaluation of measurement data – supplement 1 to the "Guide to the expression of uncertainty in measurement" – propagation of distributions using a Monte Carlo method. Joint Committee for Guides in Metrology (JCGM), Sèvres, France (2008) http://www.bipm.org/en/publications/guides/gum.html.
- [4] Toman, B. Bayesian Approaches to Calculating a Reference Value in Key Comparison Experiments, Technometrics, vol. 49, 81 – 87, 2007.

26

Thermal Conductivity Proficiency Testing Results –
Nineteen NVLAP Proficiency Testing Rounds from
1986 – 2004

-- or --

19 Years in 19 Minutes

Jeffrey Horlick and Dr. Lawrence Knab
ASTM C16.30 – Workshop 2
March 19-20, 2012

NIST National Institute of Standards and Technology •

Jeffrey Horlick
Guest Researcher
National Institute of Standards and Technology
Building 222 Room B113
100 Bureau Drive Stop 2100
Gaithersburg, MD 20899-2100

Cell: 202.812.4760
Office: 301.975.5888
Fax: 301.926.2884
E-mail: jeffrey.horlick@nist.gov

NIST National Institute of Standards and Technology •

What is NVLAP?

- A process for accreditation of testing and calibration laboratories – described in NIST Handbook 150
- Established in the U.S. Code of Federal Regulations (Title 15, Part 285) in 1976
- Administered by NIST
- Linked to NIST research units
- Based on international (ISO/IEC) standards
- Available to any qualifying laboratory (public - private)
- Fee supported
- Signatory to several international MRAs
- 850+ labs in the system today

NIST National Institute of Standards and Technology •

NVLAP Assessment of Laboratories
(Tim Rasinski is the NVLAP Program Manager for this program)

- Review of laboratory management system documentation and NVLAP assessment history

- On-site visit by NVLAP Assessors for management system and technical assessment

- Proficiency testing

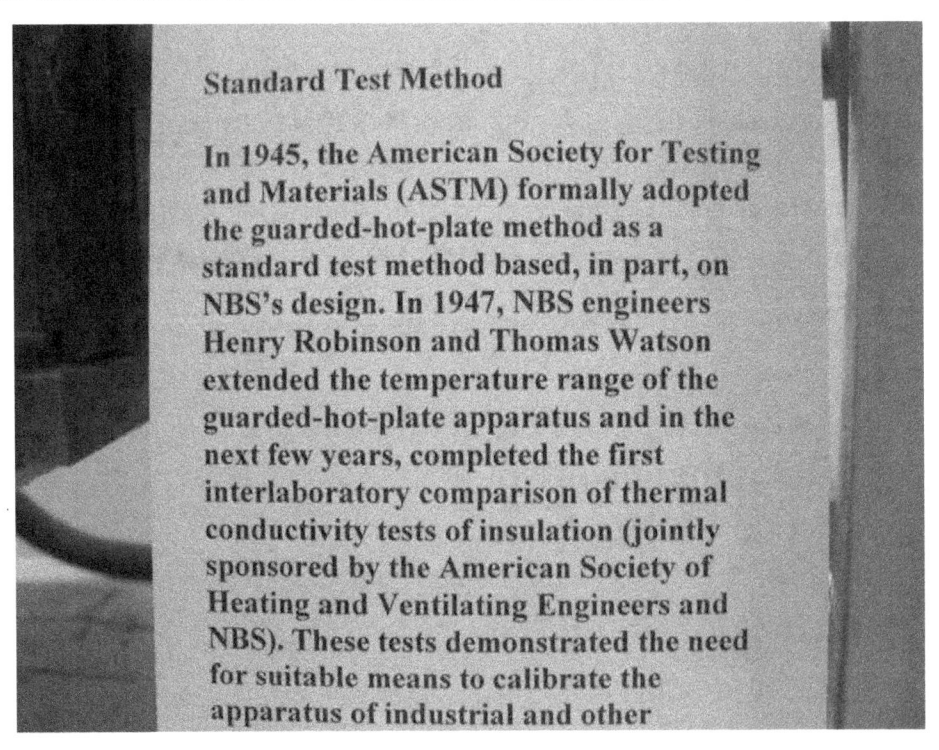

Standard Test Method

In 1945, the American Society for Testing and Materials (ASTM) formally adopted the guarded-hot-plate method as a standard test method based, in part, on NBS's design. In 1947, NBS engineers Henry Robinson and Thomas Watson extended the temperature range of the guarded-hot-plate apparatus and in the next few years, completed the first interlaboratory comparison of thermal conductivity tests of insulation (jointly sponsored by the American Society of Heating and Ventilating Engineers and NBS). These tests demonstrated the need for suitable means to calibrate the apparatus of industrial and other

Round	Year	Material Used in Each Proficiency Test Round
11	1986	1-inch thick, fibrous glass board
12	1988	1-inch thick, expanded polystyrene (EPS) board, 2.5 pcf density
13	1989	The test data for Round 13 is not included in this paper.
14	1990	0.75-inch thick, closed cell neoprene foam sheet stock
15	1991	1-inch thick, closed cell polyisocyanurate foam board with bonded faces
16	1991	0.75-inch thick, thermal insulation board composed of expanded vermiculite, binders, and fibers
17	1992	1-inch thick, unfaced, polyimide foam board
18	1992	0.5-inch thick, unfaced, rigid, closed cell PVC foam board
19	1993	2-inch thick, unfaced, rigid, cellular, polystyrene bead board (Type C Bead) (RCPS), nominal 3 pcf density
20	1993	1-inch thick, unfaced, coated one side, fibrous glass duct liner, nominal 2 pcf density
21	1994	1-inch thick, foil-faced fibrous glass board, nominal 3 pcf density
22	1995	2-inch thick, unfaced mineral wool board, nominal 6 pcf density
23	1996	1.5-inch thick, foil-faced fibrous glass duct wrap, 0.75 pcf density
24	1997	2-inch thick, unfaced fibrous glass board, 3-pcf density
25	1997	1-inch thick, open cell polyurethane foam, 2.5 pcf density
26	1999	1-inch nominal thickness, expanded polystyrene (EPS) board, 2.6 pcf density
27	2000	2-inch nominal thickness, expanded polystyrene (EPS) board, 2.6 pcf density
28	2001	combined 1-inch and 2-inch nominal thickness to make a 3-inch specimen; expanded polystyrene (EPS) board, 2.6 pcf density. In most cases, labs used the one-inch EPS test sample from previous R26 and likewise, the two inch EPS test sample from previous R27.
29	2003	1-inch nominal thickness, fibrous glass duct liner insulation, 1.6 to 1.8 pcf density
30	2004	1.5-inch nominal thickness, polyisocyanurate rigid-board insulation, 2.5 pcf density

Brief summary of possible factors affecting proficiency testing results

a) Time: this report covers 19 rounds of NVLAP Proficiency Testing: Rounds 11 through 30, 1986 through 2004 – 19 years

b) Test method standards: ASTM C177 and ASTM C518 have changed. Between 1986 and 2004 there were three revisions to ASTM C177 and five revisions to ASTM C518

c) k factor value: use of correct and consistent units, calculations, conversion factors, thickness discrepancies (see "Thickness" below); hot and cold temperatures used; temperature gradients used

d) Laboratories: 37 laboratories with differing time spans of accreditation, laboratory mix in each round, number of laboratories in each round - laboratories were research, independent, or manufacturers of insulation or thermal conductivity measuring apparatus. Not all laboratories were equipped to test all of the insulation types. Over the 19 years many changes occurred including relocation, equipment, ownership and personnel changes.

e) Density: within- specimen variation, including metering area, among- specimen variation, measurement and calculation, effect of compressible materials versus board materials; effect of light weight density on k-factor values

f) Thickness: issues of measurement of specimen thickness as tested, uniformity of thickness, test of solid and compressible materials, use of plate separation to estimate insulation testing thickness, use of spacer blocks with limited discrete thicknesses. Laboratory use of nominal versus actual thickness in calculating k-factor values. Not all laboratories could measure 1-inch, 2-inch, 3-inch thick specimens.

g) Test instructions: variation due to unclear or misinterpreted instructions; deviations from the test instructions

h) Insulation materials: fibrous glass, foam board, etc were used. Some laboratories were not familiar with the insulation supplied. Effect of off-gassing.

i) Insulation Specimen: use of insulation with facing; positioning of facing (foil, matec, coating, hard side, etc.) on hot versus cold side; variations due to uneven foil facing and insulation surfaces. Variations in metering area and guard/filler material.

j) Equipment: Laboratory-to-laboratory variation in equipment design, geometry, size, thickness capacity and use. Evolution of thermal measuring equipment over time.

k) Standard Reference Materials (SRMs): variation caused by using SRMs with thickness, density, or insulation type differing from the insulation being measured; wear of SRMs over time due to normal usage.

l) Insulation conditioning: variations in relative humidity, moisture content, and temperature

m) Guard/Filler materials around the specimen: use of different guard and filler materials for example, batting or air

223

"trimmed Grand Mean"

is the mathematic average of the k-factor values

with the outliers excluded

Percent Deviation from trimmed Grand Mean

=

100 * (laboratory value – trimmed Grand Mean)

----

trimmed Grand Mean

NIST National Institute of Standards and Technology •

ASTM C177	Round	11	12	13	14	15	16	17
Year of testing		1986	1988	1989	1990	1991	1991	1992
Trimmed Grand Mean of k-factor value		0.2304	0.2348		0.2619	0.1604	0.3844	0.3297
Standard Deviation of k-factor value		0.0043	0.0033		0.0039	0.0048	0.0135	0.0099
%CV of k-factor value		1.87	1.38		1.50	2.98	3.52	2.99
No. of labs / No. of labs outliers excluded		9 / 8	11 / 10		10 / 9	10 / 10	10 / 9	9 / 9

ASTM C518	Round	11	12	13	14	15	16	17
Year of testing		1986	1988	1989	1990	1991	1991	1992
Trimmed Grand Mean of k-factor value		0.2288	0.2354		0.2592	0.1638	0.3721	0.3227
Standard Deviation of k-factor value		0.0033	0.0034		0.0042	0.0045	0.0114	0.0070
%CV of k-factor value		1.43	1.43		1.61	2.73	3.08	2.16
No. of labs / No. of labs outliers excluded		29 / 26	27 / 27		20 / 19	21 / 21	21 / 20	20 / 20

ASTM C177	Round	11	12	13	14	15	16	17	18	19	20
Year of testing		1986	1988	1989	1990	1991	1991	1992	1992	1993	1993
Trimmed Grand Mean of k-factor value		0.2304	0.2348		0.2619	0.1604	0.3844	0.3297	0.2338	0.2369	0.2484
Standard Deviation of k-factor value		0.0043	0.0033		0.0039	0.0048	0.0135	0.0099	0.0088	0.0069	0.0063
%CV of k-factor value		1.87	1.38		1.50	2.98	3.52	2.99	3.75	2.91	2.52
No. of labs / No. of labs outliers excluded		9 / 8	11 / 10		10 / 9	10 / 10	10 / 9	9 / 9	10 / 9	10 / 9	10 / 9

ASTM C518	Round	11	12	13	14	15	16	17	18	19	20
Year of testing		1986	1988	1989	1990	1991	1991	1992	1992	1993	1993
Trimmed Grand Mean of k-factor value		0.2288	0.2354		0.2592	0.1638	0.3721	0.3227	0.2334	0.2346	0.2456
Standard Deviation of k-factor value		0.0033	0.0034		0.0042	0.0045	0.0114	0.0070	0.0116	0.0036	0.0084
%CV of k-factor value		1.43	1.43		1.61	2.73	3.08	2.16	4.99	1.54	3.41
No. of labs / No. of labs outliers excluded		29 / 26	27 / 27		20 / 19	21 / 21	21 / 20	20 / 20	21 / 21	21 / 20	21 / 21

ASTM C177	Round	21	22	23	24	25	26	27	28	29	30
Year of testing		1994	1995	1996	1997	1997	1999	2000	2001	2003	2004
Trimmed Grand Mean of k-factor value		0.2236	0.2304	0.2673	0.2283	0.2562	0.2278	0.2301	0.2325	0.2432	0.1779
Standard Deviation of k-factor value		0.0042	0.0061	0.0177	0.0032	0.0051	0.0006	0.0024	0.0036	0.0042	0.0042
%CV of k-factor value		1.88	2.63	6.61	1.40	2.00	0.28	1.03	1.56	1.75	2.36
No. of labs / No. of labs outliers excluded		8 / 8	8 / 8	8 / 8	7 / 6	7 / 7	7 / 5	7 / 6	7 / 7	7 / 6	5 / 5

ASTM C518	Round	21	22	23	24	25	26	27	28	29	30
Year of testing		1994	1995	1996	1997	1997	1999	2000	2001	2003	2004
Trimmed Grand Mean of k-factor value		0.2230	0.2328	0.2745	0.2291	0.2548	0.2291	0.2301	0.2335	0.2408	0.1814
Standard Deviation of k-factor value		0.0050	0.0039	0.0132	0.0042	0.0024	0.0020	0.0029	0.0039	0.0024	0.0020
%CV of k-factor value		2.22	1.67	4.82	1.81	0.93	0.88	1.28	1.66	0.98	1.11
No. of labs / No. of labs outliers excluded		20 / 18	14 / 14	14 / 14	15 / 14	15 / 12	14 / 11	12 / 11	10 / 10	10 / 10	8 / 8

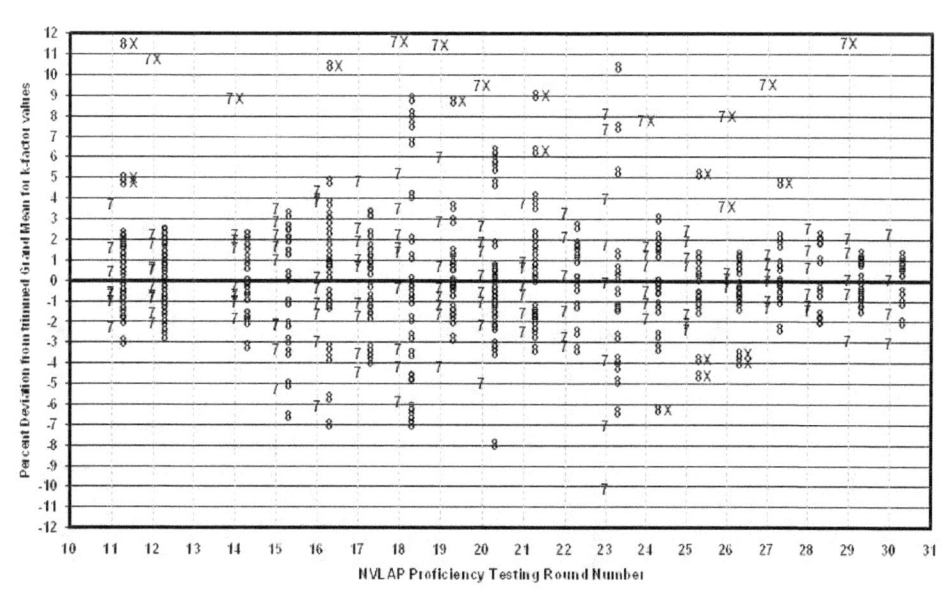

Thermal Insulation LAP
ASTM C177 and ASTM C518 using trimmed Grand Mean

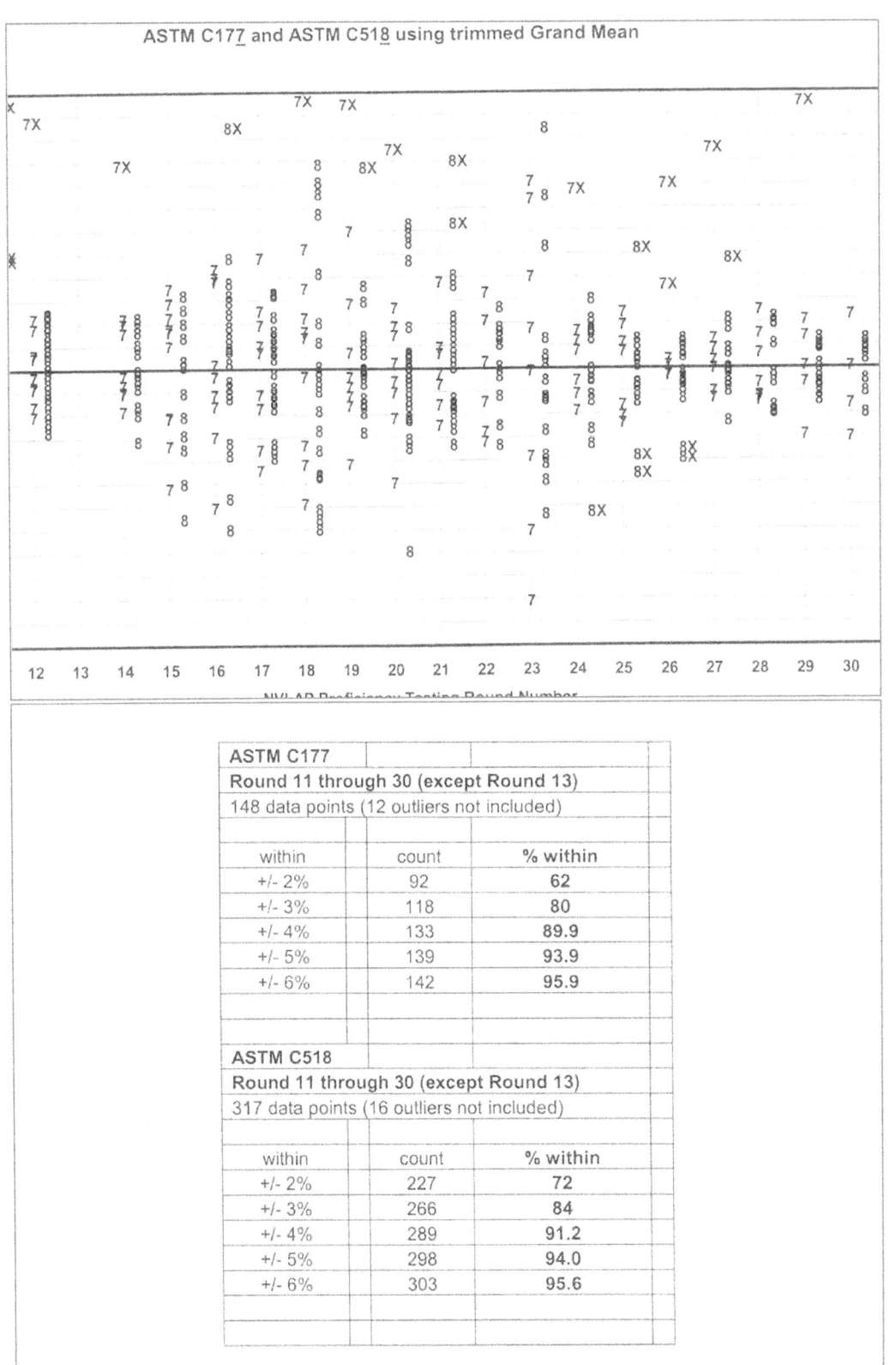

ASTM C177 and ASTM C518 using trimmed Grand Mean

ASTM C177		
Round 11 through 30 (except Round 13)		
148 data points (12 outliers not included)		
within	count	% within
+/- 2%	92	62
+/- 3%	118	80
+/- 4%	133	89.9
+/- 5%	139	93.9
+/- 6%	142	95.9
ASTM C518		
Round 11 through 30 (except Round 13)		
317 data points (16 outliers not included)		
within	count	% within
+/- 2%	227	72
+/- 3%	266	84
+/- 4%	289	91.2
+/- 5%	298	94.0
+/- 6%	303	95.6

226

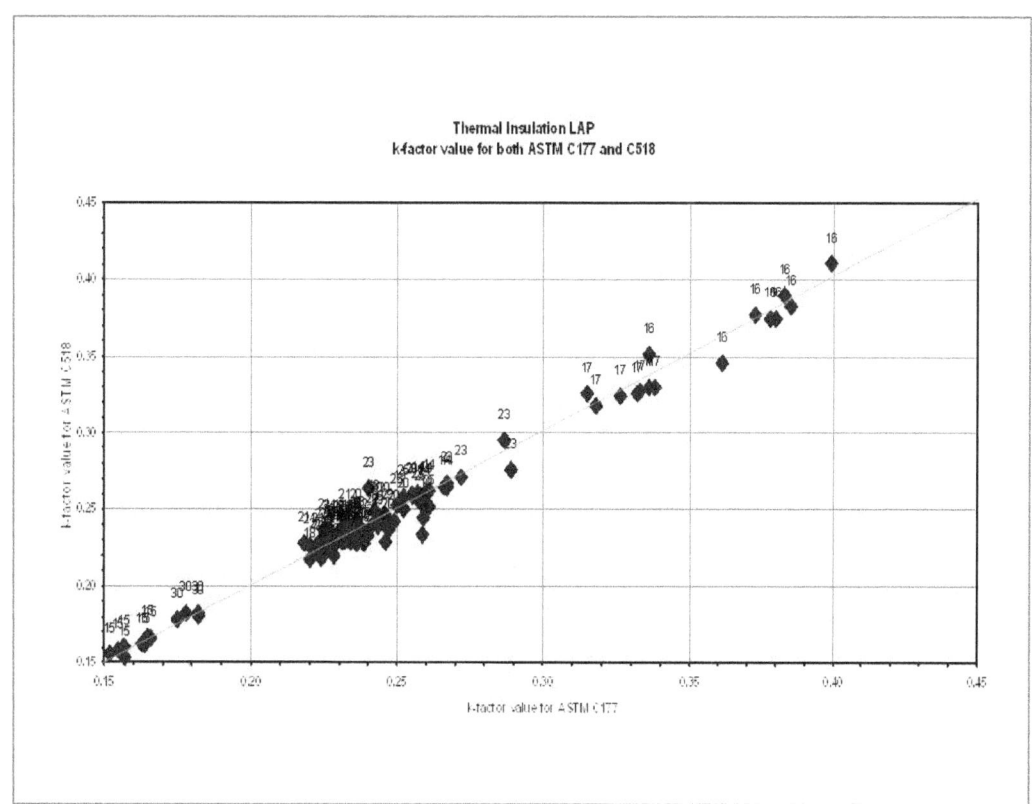

Thermal Insulation LAP
k-factor value for both ASTM C 177 and C518

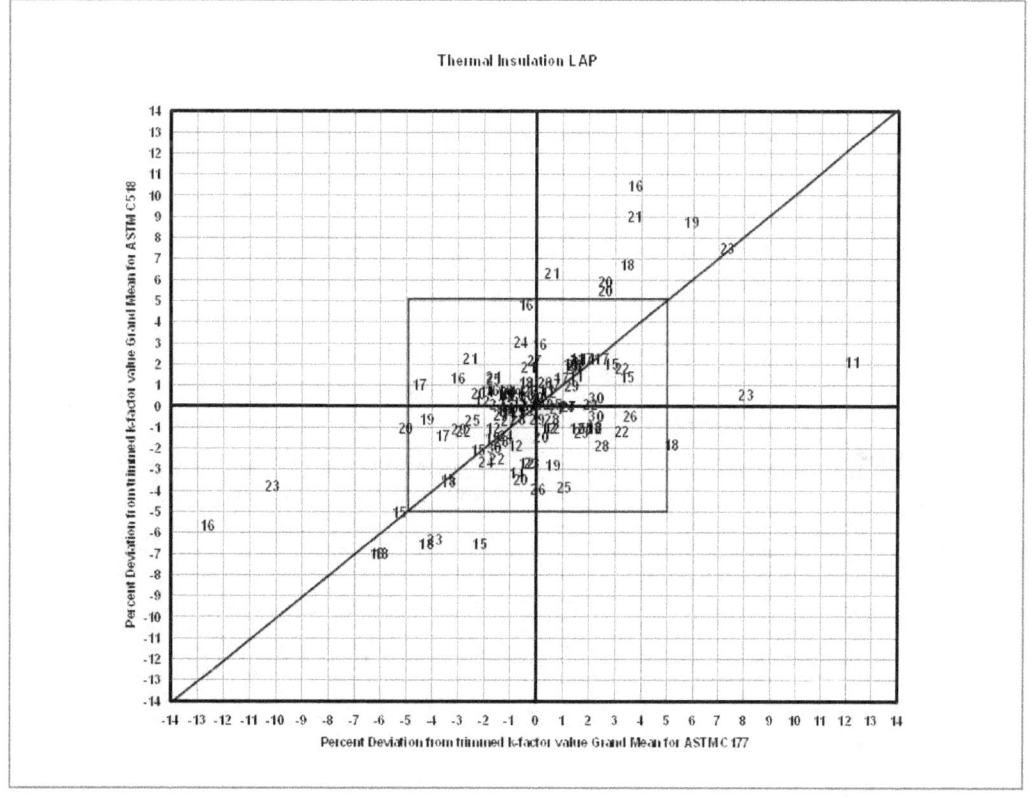

Thermal Insulation LAP

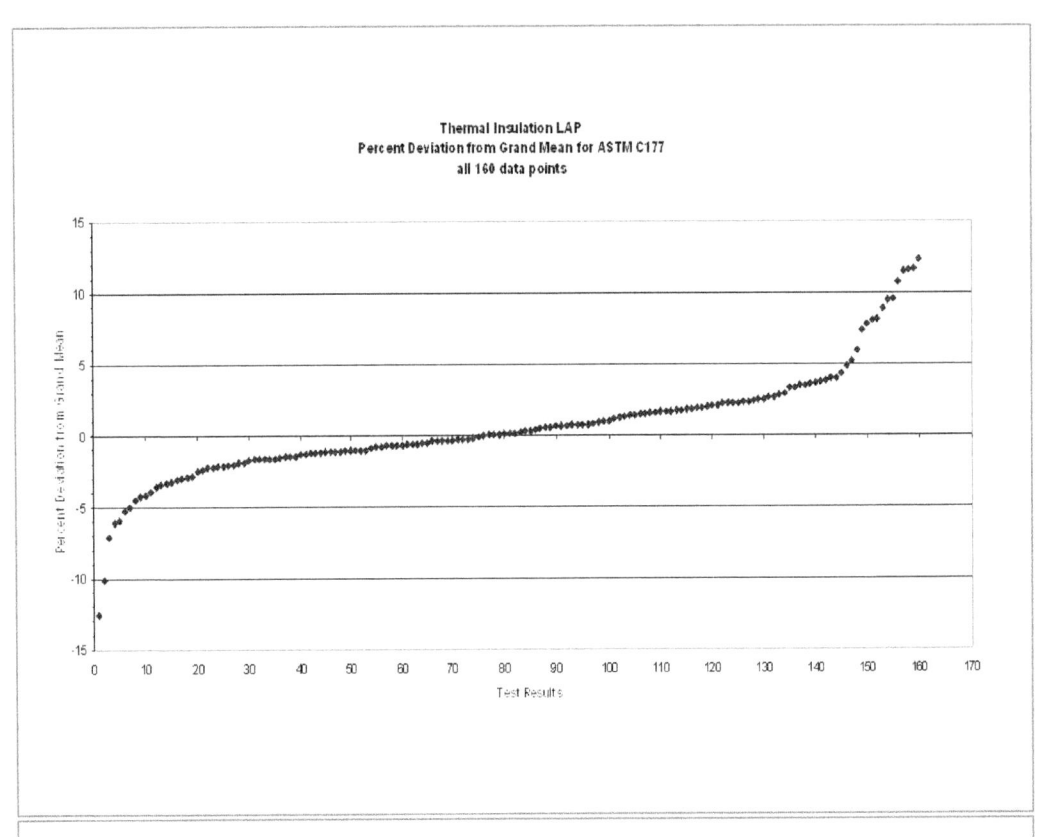

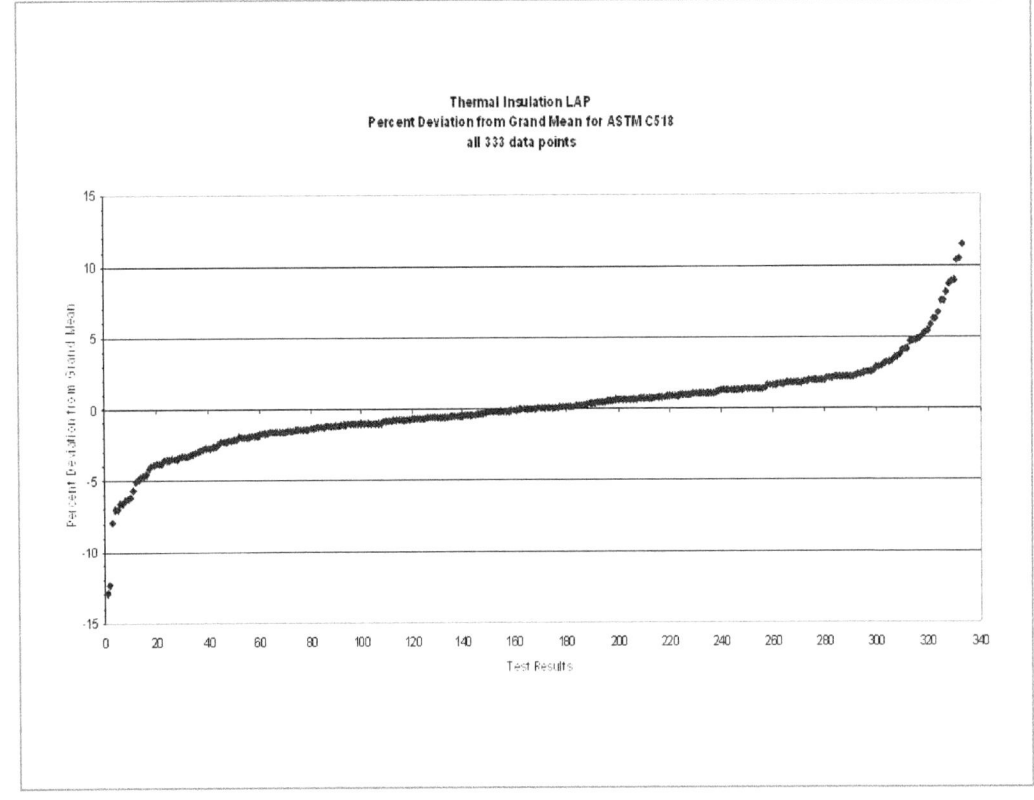

228

Next

- Data from NVLAP Proficiency Testing have been used in the development of precision and bias statements for ASTM C177 and C518. The data analyzed in this presentation has been made available to ASTM Subcommittee C16.30.

- Reprints are available from Jeffrey Horlick.

- NVLAP will continue to conduct proficiency testing for thermal conductivity when a contract proficiency testing provider is found. The last round was #31 in February 2010.

EURIMA
European Mineral Wool
Association

High Temperature Thermal
Conductivity Measurment
Comparisons

EURIMA Comparison 1996

The tests were performed on
individual specimens taken from a
selected lot of high density Rockwool
boards resulting in a density range
for the specimens from 134 kg/m^3 to
152 kg/m^3.

Measured Thermal Conductivities [mW/(m*K)]

1996					
Mean temp	100 C	200 C	300 C	400 C	500 C
H	48.6	63	79.6	99.2	122.4
I	39.8	51.8	65.1	80.7	99.7
K	43	57	72	89	108
L	43	56	72	91	112
M	43.1	56.4	71.3	87.8	
N	45	58.1	73.6	91.7	112
O	40.2	51.6	65.2	81.1	99.2
P	44	57	72.2	90.8	114
Q	39.4	46.1	57.8	74.6	
Mean	42.9	55.2	69.9	87.3	109.6
Max deviation from mean %	- 8	- 16	-17	- 14	- 8

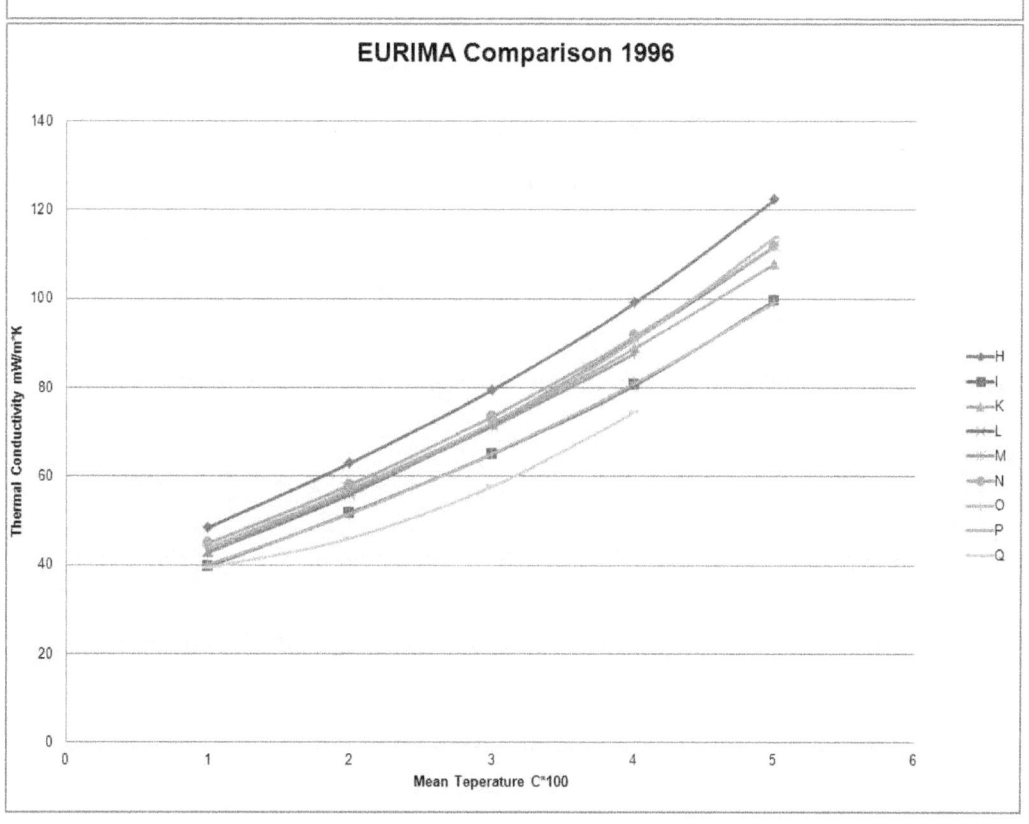

EURIMA Comparison 2011

Contrary to the 1996 comparison the 2011 one used the same specimen with the limitation due to different plate sizes for the used equipment. The specimen was circulated to the laboratories in order of decreasing plate dimensions. The specimen was a ISOFRAX blanket measured at a thickness of 38 mm. The densities stated by the laboratories varied from 124 kg/m^3 to 135 kg/m^3

Measured Thermal Conductivities [mW/(m*K)]

			2011		
Mean Temperature	100 C	200 C	300 C	400 C	500 C
A	46.3	65.3	88.8	120.0	161.6
B	49.0	66.0	89.0	119.0	159.0
C	49.9	68.2	91.1	120.8	159.0
E		67.6	90.0	119.0	156.7
F	47.9	67.6	94.8	128.4	171.6
G		62.0	83.7	111.4	145.1
Mean	48.3	66.1	89.6	119.8	158.8
Max deviation from mean %	4.1	6.2	6.6	7.0	8.6

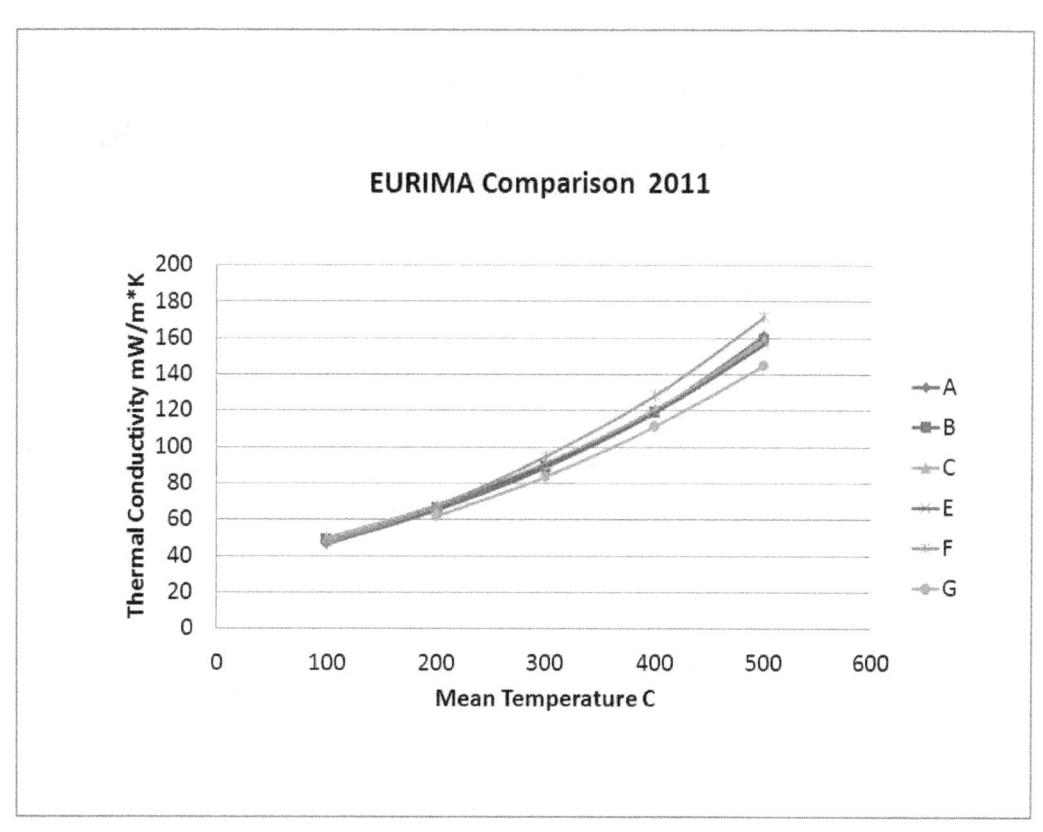

2nd Operators Workshop on High-Temperature Guarded-Hot-Plate and Pipe Measurements

High-Temperature Thermal Insulation Industry Needs

Thomas Whitaker
Chairman, ASTM Committee C16 on Thermal Insulation

C16 Round Robin
C335 Pipe Test

- The goal of this RR is to evaluate equipment differences. The participating laboratories all have different designs.

- In previous RR there have been issues with the material chosen for the RR. There were "fit issues" when Calcium Silicate was circulated. When we used Mineral Wool, the pieces were characterized by dimensions and weight but there were no assurances the pieces were identical.

- There participants have agreed to providing construction details about the apparatus
- The temperature has been restricted to 300C as the upper limit due to limits of the participants.
- The participants have agreed to supply detailed measurements not just the "steady state" data.

Issues

- Drift is an issue and has caused some discussion. For both the test methods we are looking at in the workshop, the "definition" of steady state is change of <0.5% and not monotonically increasing or decreasing. If an operator does not have computer controlled equipment that collects data and proceeds to the next point, he can see a longer term, slow drift. I have allowed my apparatus to sit at one temperature and plot the measured conductivity. Can meet the definition of "steady state" in about 1 day, but after 5 days, the answer is different and sometimes by more than 1%. My apparatus sets in a room measuring about 4m x 5m. If I plot measured conductivity over time, I can pinpoint when the night crew comes in to empty the trash containers. That visit could trigger a change in the slope and provide an inflection point that breaks the "monotonically changing" rule.

- Applies to both Pipe and Flat test methods. At higher temperatures, lots of things change. If the test sample has some organic or combustible component, that can shift the measured conductivity.

- Some materials will have a reaction somewhat quickly and fit within the definition of "steady state". Others a delayed reaction that can cause a shift, but it occurs slowly and may or may not be complete when the computer considers the apparatus to be at "steady state".

Questions & Discussion

2nd Operators Workshop on High-Temperature Guarded-Hot-Plate and Pipe Measurements

ASTM 16:30 Thermal Measurements Subcommittee
Test Methods C 177 and C 335

Experiment Design (DEX) of Round Robin

Jim Filliben, NIST

03/20/12 (Tuesday): 1:50-2:30
224/B245

Bob Zarr
jjfzarrtalk032012.pptx
Handout Notes

The deliverable of this talk is

1. to describe a structured and generic 2-part framework that has proved to be useful and insightful for scientific problem-solving at NIST; and

2. to apply that structure to the problem of assessing whether a set of laboratories are equivalent wrt their thermal conductivity measurements.

Outline

1. DEX/Stat Framework
2. Questions
3. DEX Problem Classification
4. DEX Problem Translation
5. DEX Principles & Techniques
6. DEX Criteria & Interlab Designs

Slide 1

Physics
Plutonium Troubleshooting (SURF)
Am 241/243 Peal Deconv. Alg. Acc.
Cesium 137 Detection
Efficiency of Gamma Ray Emitters
Remote Radiation Detection (SURF)
Sonoluminescent Light Intens.(SURF)
ASP (Adv. Spectrosc. Portal) Monitor.
PRD (Personal Radiation Detectors)
Maritime Radiation Detectors
Soil Leeching Seq. Extraction Prot.

NIST

Material Science
MALDI TOF Spectrometry
Nanocantilever Atomic Force Mic.
Dental Polysac Adhesion
Bio Knee Cartilage Regeneration
Ceramic Machining Strength
Comb. Chemistry Tape Peel

Chemistry
Carbon Nanotube Water Pollution
SRM 2396: DNA Base Biomarkers
Gate Dialectrics: SiO2 HRTEM Error
Microarray Sensors for Toxic Gas
DHS: Bio-Agent Detection
Radiocarbon C14 Albuq. CO Pollut.
(Cu-AU) 3D Nanoscale Chem. Imaging
Dual Rotor Turbin Fluid Flow
SO2 Permeation Tube Mass Loss
KC (Key Comparison) Fluid Flow

Elect. & Elect. Eng.
OLES: Bullet Proof Vest Reliability
Eddy Current Probe
IACP/OLES: Safety/Speed Devices Acceptance Samp.
DAC (Digital-to-Analog Converter) Calibration
OLES: Firefighter Infrared Imaging Devices
OLES: Metal Detector Acceptance Sampling

Building & Fire Research
World Trade Center FEA Core Damage
Cigarette Ignition Propensity
FHWA Highway Concrete Strength (COST)
Tall Building Deflection Safety Codes
HHS CONTAM Home Pollution Dissemination
Solar Sphere Testing of Polymeric Sealants
Optimization of Hot Plate Gap Parameters
Interlab: Thermal Hot Plate Conductivity
Tomographic Flow Detection in Polymer-Bonded Concrete
HUD Lead Paint Test Kit Accuracy
HUD Lead Paint Extraction
Hospital Energy Consumption
Evaluating Strategies for Fire Safety
Paint Peel Strength
Aerosol Spray Flow Rates
Asphalt Roofing Vertical Peel Testing
Remote Detection of Pre-Mold Moisture in Building Mats.
WTC FDS (Fire Dynamics Simulator) Sensitivity
WTC FDS Validation
WTC Impact Sensitivity
WTC FEA Insulation-on-Steel Thermal Propagation
WTC Structural Sensitivity

Manufacturing Eng.
Scatterfield Microscopy
Genetic Alg. for Machine Tooling
SMS: Smart Machining System
NIJ/OLES: Forensic Imaging of Gun Casings

Information Tech.
Abilene Network Internet Congestion Modeling
Cloud Computing Resource Allocators
Accelerated Testing of Compact Discs
Bio-Cell Imaging Segmentation Algorithms
RAVE Visualization Facility Calibration
Biometrics(2): Iris, 3D Fingerprints
Ct Scanner Dosage Analysis

Slide 2

Physics
Plutonium Troubleshooting (SURF)
Am 241/243 Peal Deconv. Alg. Acc.
Cesium 137 Detection
Efficiency of Gamma Ray Emitters
Remote Radiation Detection (SURF)
Sonoluminescent Light Intens.(SURF)
ASP (Adv. Spectrosc. Portal) Monitor.
PRD (Personal Radiation Detectors)
Maritime Radiation Detectors
Soil Leeching Seq. Extraction

NIST

Material Science
MALDI TOF Spectrometry
Nanocantilever Atomic Force Mic.
Dental Polysac Adhesion
Bio Knee Cartilage Regeneration
Ceramic Machining Strength

Chemistry
Carbon Nanotube Water Pollution
SRM 2396: DNA Base Biomarkers
Gate Dialectrics: SiO2 HRTEM Error
Microarray Sensors for Toxic Gas
DHS: Bio-Agent Detection
Radiocarbon C14 Albuq. CO Pollut.
(Cu-AU) 3D Nanoscale Chem. Imaging
Dual Rotor Turbin Fluid Flow
SO2 Permeation Tube Mass Loss
Comparison) Fluid Flow

Elect. & El
OLES: Bullet Proof Vest Reli
Eddy Current Probe
IACP/OLES: Safety/Speed D
DAC (Digital-to-Analog Conv
OLES: Firefighter Infrared Im
OLES: Metal Detector Accep

All of these problems were addressed by the structured approach to be described in this talk ...

ire Research
Damage
ngth (COST)
y Codes
Dissemination
eric Sealants
Optimization of Hot Plate Gap Parameters
Interlab: Thermal Hot Plate Conductivity
Tomographic Flow Detection in Polymer-Bonded Concrete
HUD Lead Paint Test Kit Accuracy
HUD Lead Paint Extraction
Hospital Energy Consumption
Evaluating Strategies for Fire Safety
Paint Peel Strength
Aerosol Spray Flow Rates
Asphalt Roofing Vertical Peel Testing
Remote Detection of Pre-Mold Moisture in Building Mats.
WTC FDS (Fire Dynamics Simulator) Sensitivity
WTC FDS Validation
WTC Impact Sensitivity
WTC FEA Insulation-on-Steel Thermal Propagation
WTC Structural Sensitivity

Manufacturing Eng.
Scatterfield Microscopy
Genetic Alg. for Machine Tooling
SMS: Smart Machining System
NIJ/OLES: Forensic Imaging of Gun Casings

Information Tech.
Accelerated Testing of Compact Discs
Abilene Network Internet Congestion Modeling
Cloud Computing Resource Allocators
Bio-Cell Imaging Segmentation Algorithms
RAVE Visualization Facility Calibration
Biometrics(2): Iris, 3D Fingerprints
Ct Scanner Dosage Analysis

NIST Inter-Governmental Collaborations:

Congress
World Trade Center

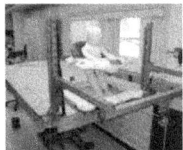
CPSC
Consumer Testing

GSA
LADAR Building Inspection

Bullet proof Vests

Radiation Detection

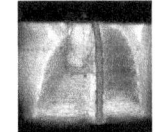
Lung Tumor Metrology

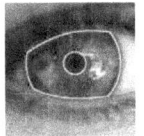
Iris-scan Identification

Food Gamma-ray Spectrometry

Gulf Oil Spill

10

1. DEX/Stat Framework

General Problem-Solving <u>*Framework*</u>

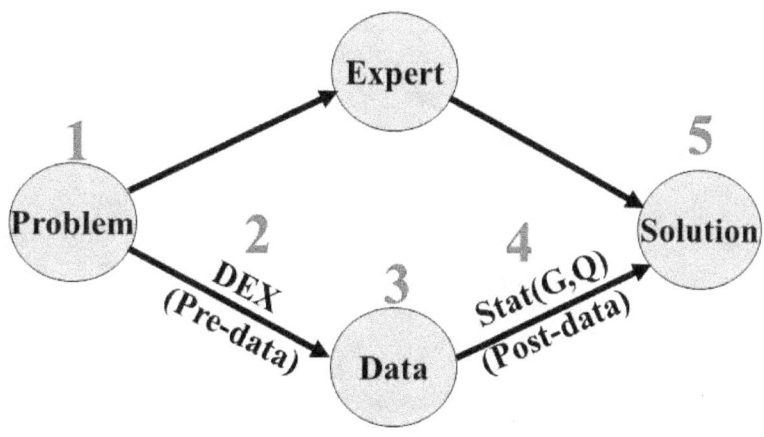

5-Step Data-Based Generic Framework

General Problem-Solving <u>Framework</u>

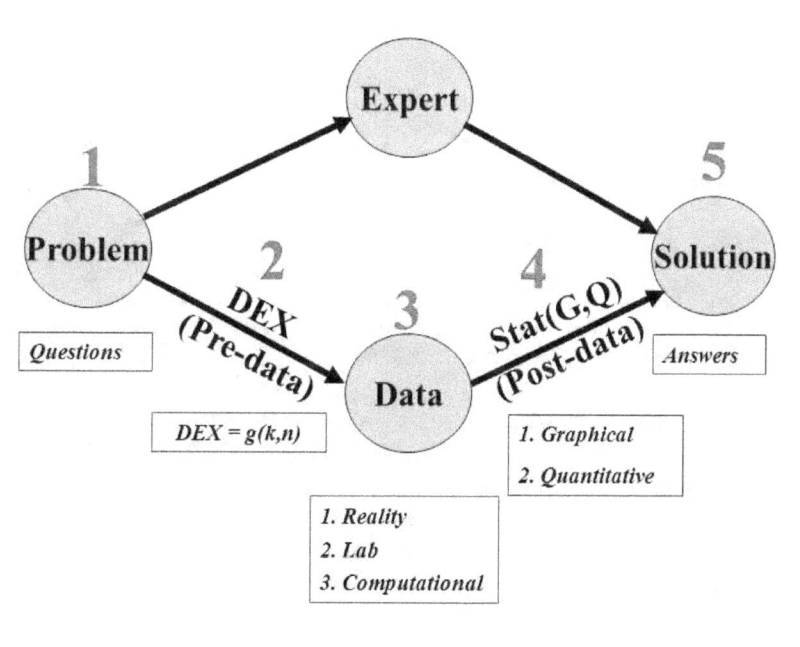

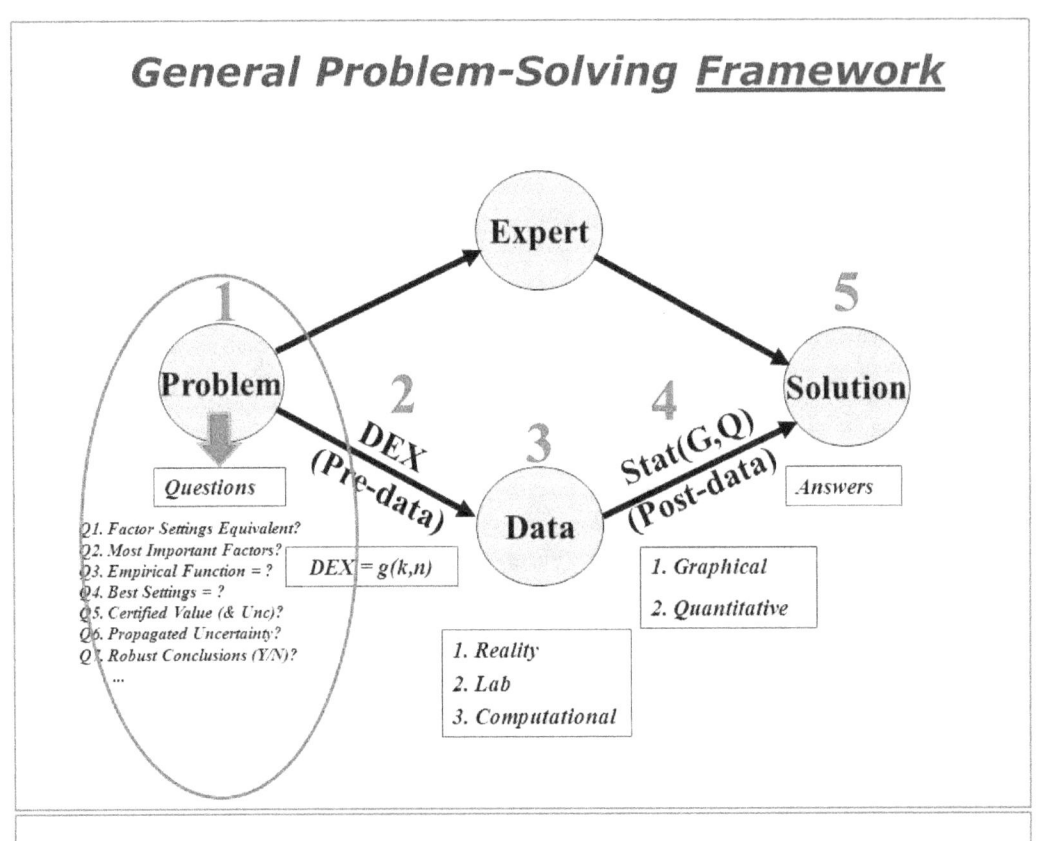

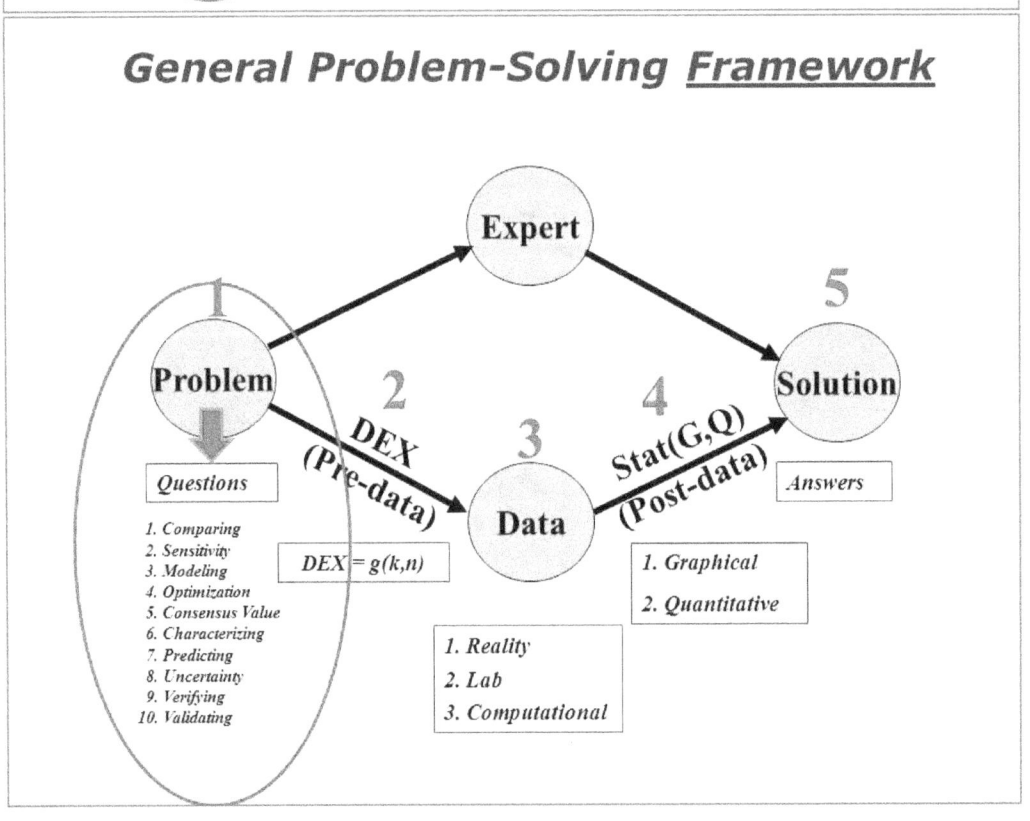

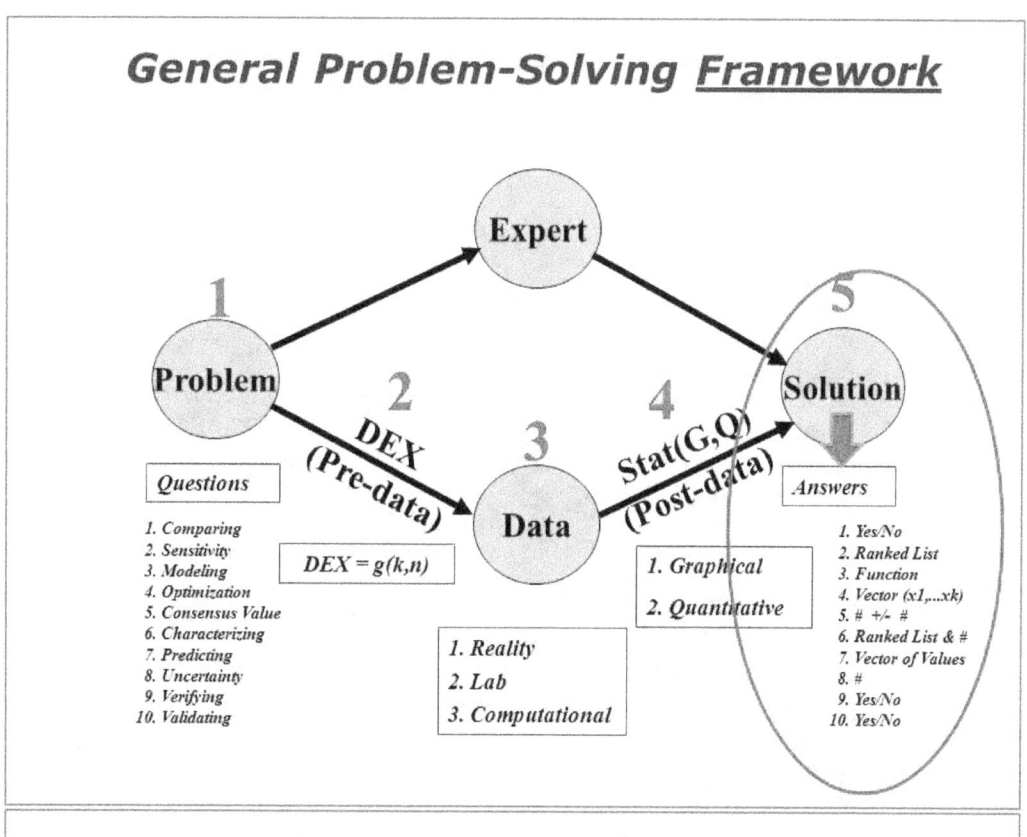

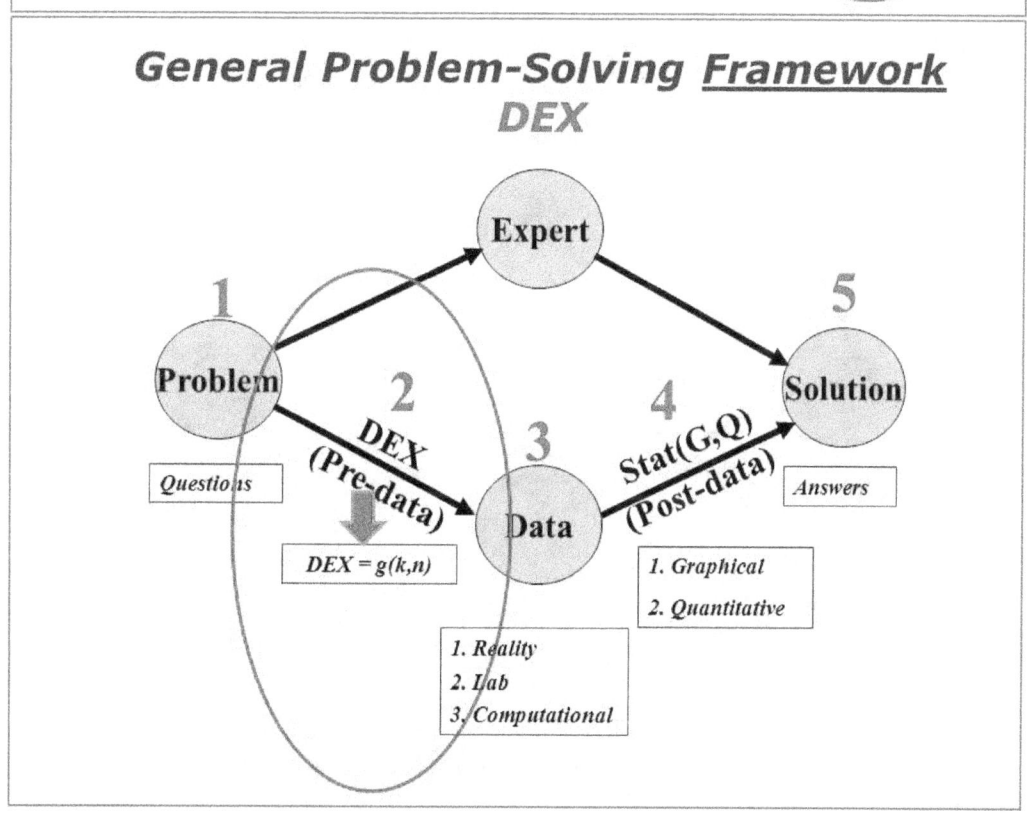

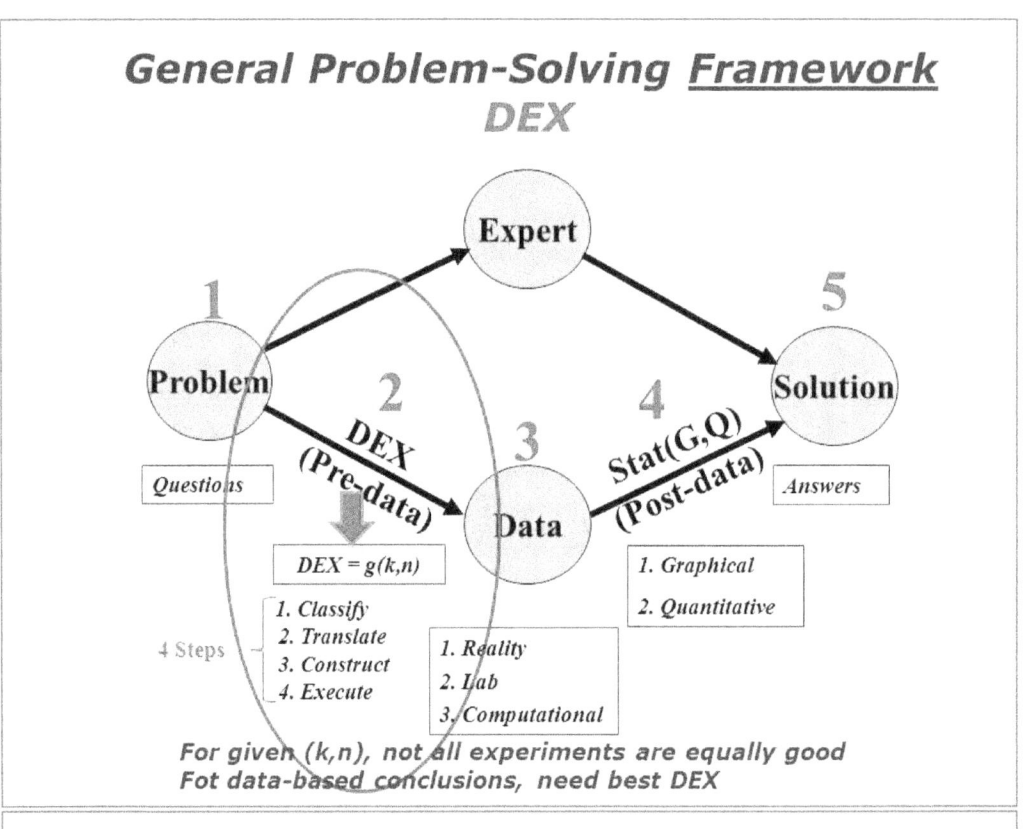

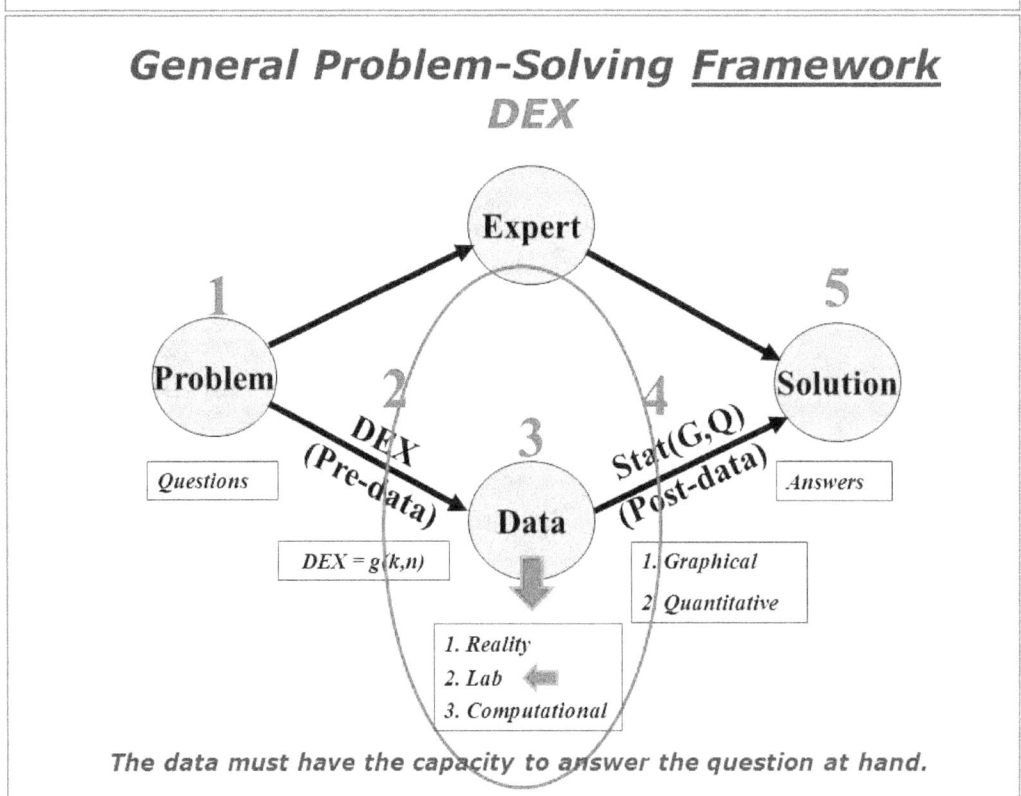

244

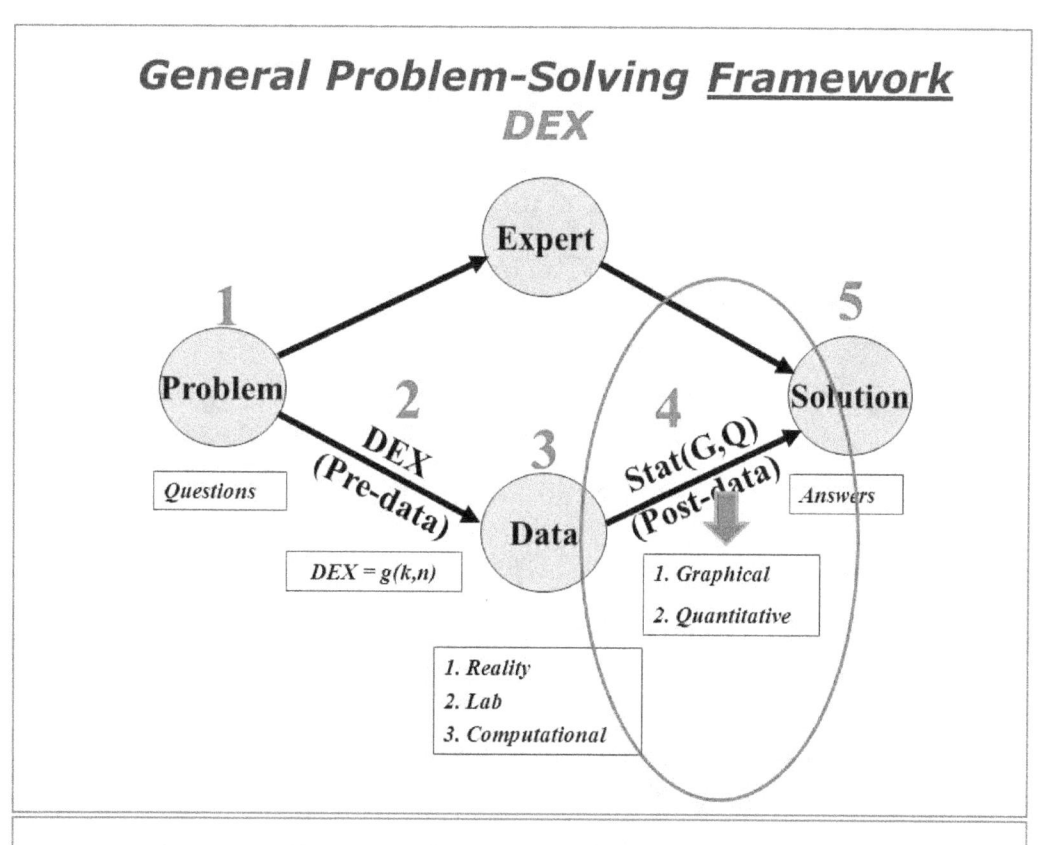

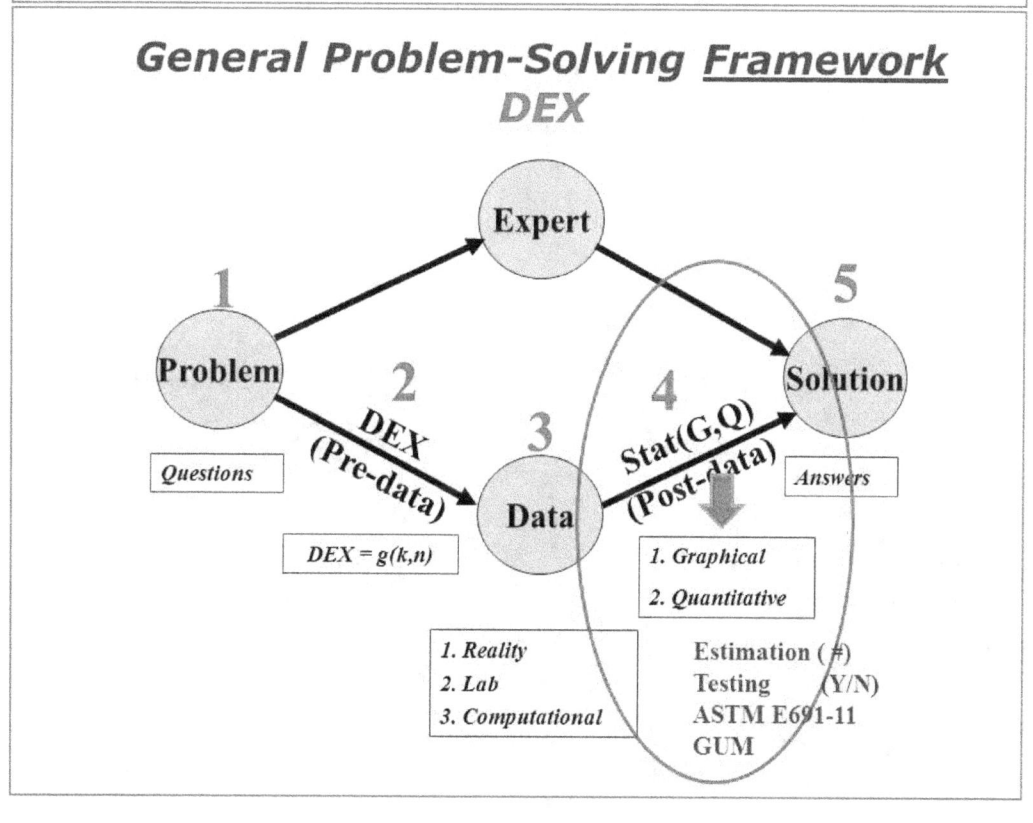

2. Questions

General Problem-Solving *Framework*

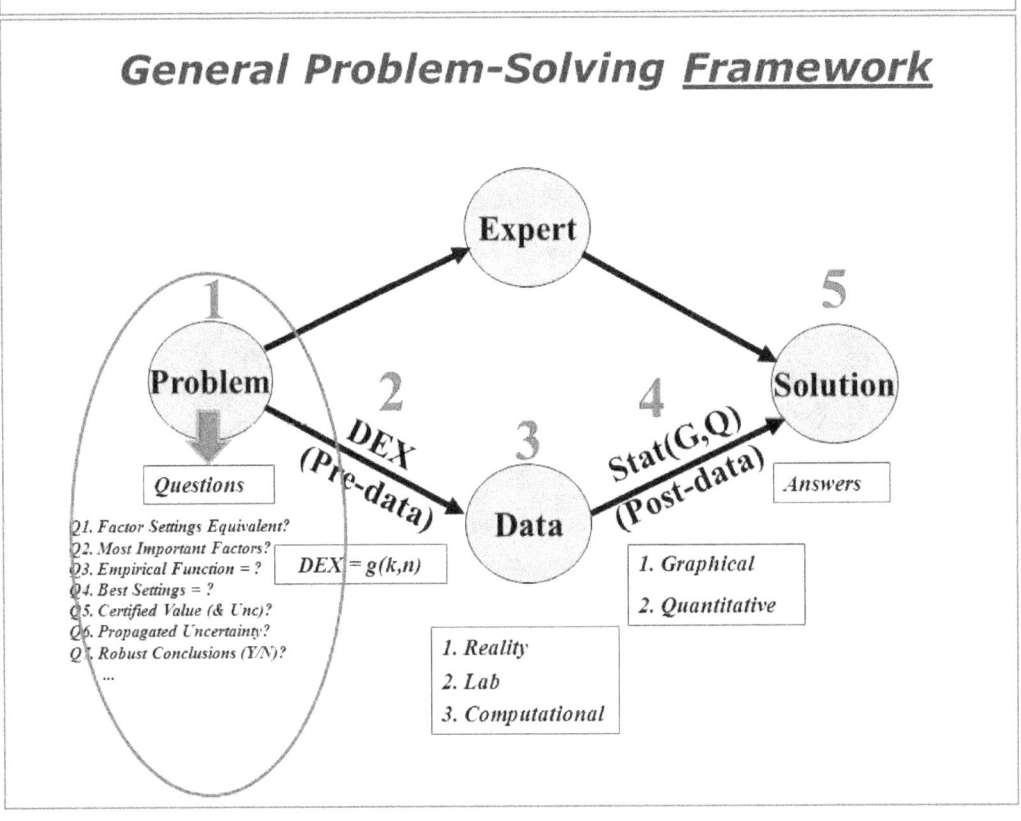

Within-Lab Questions

For my lab ...

Q1. What is the uncertainty of my thermal cond. (TC) measurements?

Q2. Does my TC value change with material?

Q3. Does the unc(TC value) change with material, or temp, or ...?

Q3. What are the most important factors affecting the uncertainty, bias, and precision of my TC measurements?

Q4. What are the optimal settings for all relevant factors?

Q5. What is a good predictive model, that describes my TC measurements? Is such a model good enough for simulation?

Q6. Are the above answers robust over all relevant other ("robustness") factors that may effect my system?

Q7. What is the temperature scope ("range") of my TC readings?

Q8. To test any of the above, how many observations do I (stat) need?

Q9. To test any of the above, how many observations can I afford?

Q10. Will my sample (of readings, of materials) be representative of the population?

Q11. What is the population? (of readings, of materials, of ...)

Q12. Are my TC readings in stat control?

Q13. Does ground truth exist?

Within-Lab Questions

For my lab ...

Q1. What is the uncertainty of my thermal cond. (TC) measurements? # +- #

Q2. Does my TC value change with material? Y/N

Q3. Does the unc(TC value) change with material, or temp, or ...? Y/N

Q3. What are the most important factors affecting the uncertainty, bias, and precision of my TC measurements? List

Q4. What are the optimal settings for all relevant factors? Vector

Q5. What is a good predictive model, that describes my TC measurements? Is such a model good enough for simulation? Function

Q6. Are the above answers robust over all relevant other ("robustness") factors that may effect my system? Y/N

Q7. What is the temperature scope ("range") of my TC readings? [#,#]

Q8. To test any of the above, how many observations do I (stat) need? #

Q9. To test any of the above, how many observations can I afford? #

Q10. Will my sample (of readings, of materials) be representative of the population? Y/N

Q11. What is the population? (of readings, of materials, of ...) {...}

Q12. Are my TC readings in stat control? Y/N

Q13. Does ground truth exist? Y/N

Between-Lab Questions

For labs 1 and 2 (and beyond) ...

Q1.	Are the 2 labs equivalent?	Y/N
Q2.	What does "equivalent" mean?	<discuss>
Q3.	Does our equivalence conclusion change with material, or temp, or ...? (interactions?) (robustness)	Y/N
Q4.	What are the most important factors affecting the our equivalence conclusion?	List
Q5.	What is a consensus TC value valid over both labs?	# +- #
Q6.	Are the devices used by the 2 labs equivalent?	Y/N
Q7.	If the labs are not equivalent, what are the correction factors which map one lab into another?	# +- #
Q8.	What is the temperature scope ("range") for my equivalence conclusion?	[#,#]
Q9.	To test equivalence, how many observations do I (stat) need?	#
Q10.	To test equivalence, how many observations can I afford ($/time)?	($, days)
Q11.	To test equivalence, how many factors should I vary?	#
Q12.	To test equivalence, How many levels of a factor (e.g., Temp) should I use? What such levels?	#
Q13.	What is the uncertainty for TC measurements in each lab?	# +- #
Q14.	Do I have data-based estimates of such uncertainties?	Y/N

3. DEX Problem Classification

General Problem-Solving *Framework*
DEX

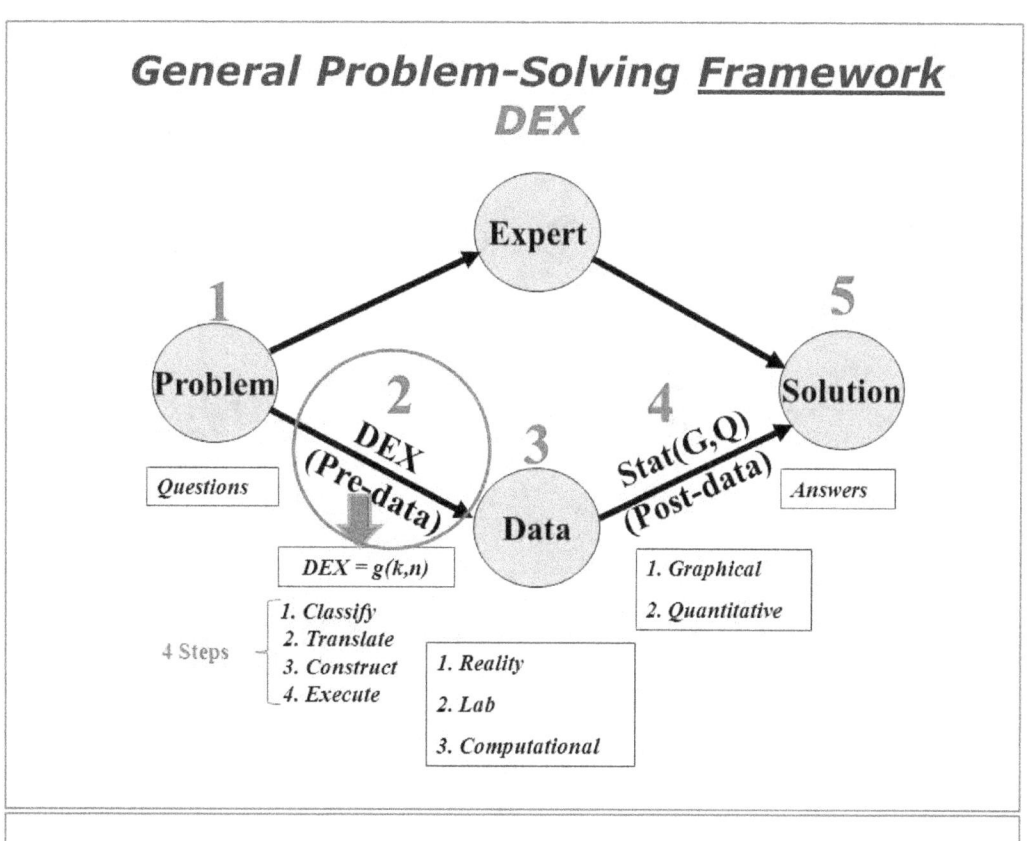

General Problem-Solving *Framework*
DEX Step *1*: *Classify*

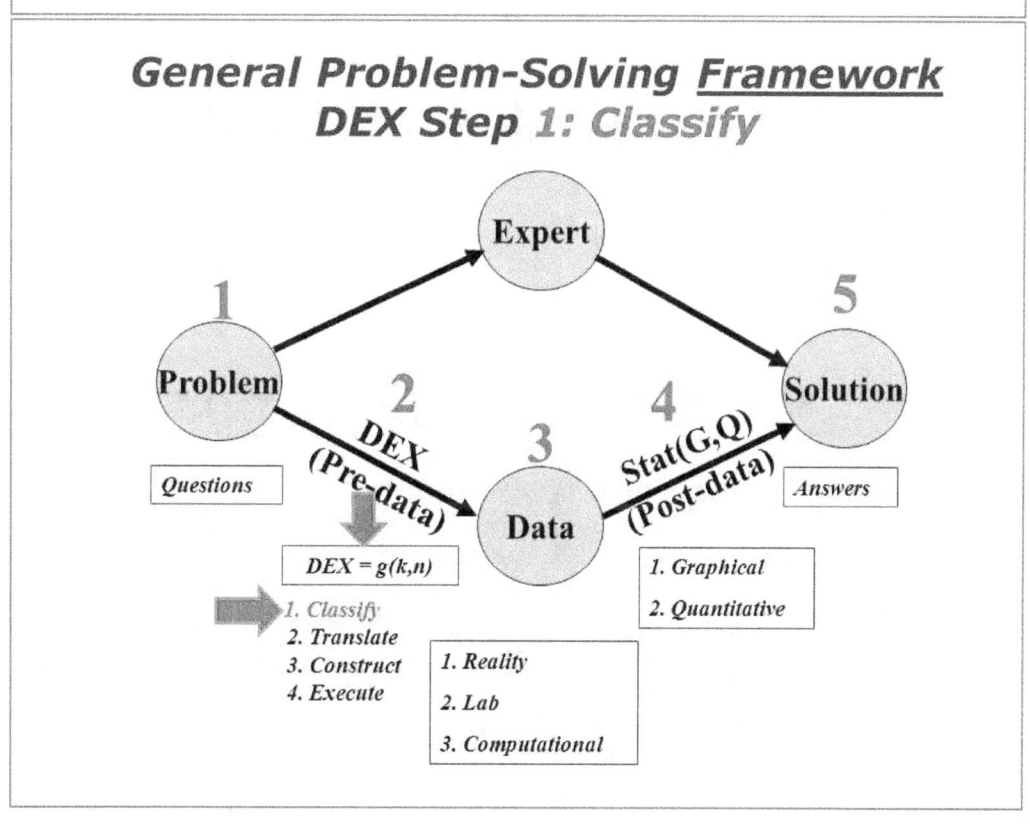

1. Problem Classification

The Starting Point: The Generic System Model:

$$Y = f(X1, X2, X3, \ldots, Xk)$$

1. Problem Classification

Is this Factor Significant?   *Most Important Factors?*

$$Y = f(X1, X2, X3, \ldots, Xk)$$

Good Approximating Function?   *Best Settings of the k Factors?*

➔ *4 (Common) Problem Categories*

1. Problem Classification

Is this Factor Significant?
1. Comparative/Robust

Most Important Factors?
2. Screening/Sensitivity

$$Y = f(X1, X2, X3, ..., Xk)$$

Good Approximating Function?
3. Modeling/Regression

Best Settings of the k Factors?
4. Optimization

---

1. Problem Classification

Is this Factor Significant?
1.   Comparative/Robust
Comp. Rand., Rand. Bl., Lat Sq, Tag. PD

Most Important Factors?
2. Screening/Sensitivity
2^k, 2^(k-p), Taguchi

$$Y = f(X1, X2, X3, ..., Xk)$$

Good Approximating Function?
3. Modeling/Regression
Randomized, Box-Behnken, XO

Best Settings of the k Factors?
4. Optimization
Resp. Surf., CCD, BB

1. Problem Classification

Is this Factor Significant?
1. Comparative/Robust
Comp. Rand., Rand. Bl., Lat Sq, Tag. PD

Most Important Factors?
2. Screening/Sensitivity
2^k, 2^(k-p), Taguchi

$$Y = f(X1, X2, X3, \ldots, Xk)$$

Good Approximating Function?
3. Modeling/Regression
Randomized, Box-Behnken, XO

Best Settings of the k Factors?
4. Optimization
Resp. Surf., CCD, BB

1. Problem Classification

1. Comparative	2. Screening/Sensitivity
Focus: 1 primary factor	Focus: all factors
Q1. Does that factor have an effect (Y/N)?	Q1. Most important factors (ranked list)
Q2. If yes, then best setting for that that factor = ? (vector)	Q2. Best settings (vector)
Constraint: Want conclusions to be robust over all other factors	Q3. Good model (function)
Designs: CRD, RBD, LSqD, TPD	Designs: 2^kD, 2^{k-p}D, TD
BHH, Ch. 4	BHH, Ch. 5-6
3. Regression	**4. Optimization**
Focus: all factors	Focus: all factors
Q1. Good model (function)	Q1. Best settings (vector)
Continuous factors	Continuous factors
Designs: BBD, XOD	Designs: RSD, CD, BBD
BHH, Ch. 10-11	BHH, Ch. 12

1. Problem Classification

1. Comparative Focus: 1 primary factor Q1. Does that factor have an effect (Y/N)? Q2. If yes, then best setting for that that factor = ? (vector) Constraint: Want conclusions to be robust over all other factors Designs: CRD, RBD, LSqD, TPD BHH, Ch. 4	**2. Screening/Sensitivity** Focus: all factors Q1. Most important factors (ranked list) Q2. Best settings (vector) Q3. Good model (function) Designs: 2^kD, $2^{k-p}D$, TD BHH, Ch. 5-6
3. Regression Focus: all factors Q1. Good model (function) Continuous factors Designs: BBD, XOD BHH, Ch. 10-11	**4. Optimization** Focus: all factors Q1. Best settings (vector) Continuous factors Designs: RSD, CD, BBD BHH, Ch. 12

Interlab & KCs:

WTC Sensitivity Analysis

Factors Under Study (k):
1. Flight Speed
2. Flight Impact Location (Vertical)
3. Flight Impact Location (Horizontal)
4. Engine Assignment Set
5. Engine Strength
6. Engine Failure Strain
7. Engine Strain Rate Effects
8. Perimeter Column Strength
9. Perimeter Column Failure Strain
10. Perimeter Column Strain Rate Effects
11. FEA Model Erosion Parameter
12. FEA Contact Parameter
13. FEA Friction Coefficient

Affordable Number of Runs: n < 50

$$DEX = g(k, n)$$

$$(k = 13,\ n < 50)$$

(Design and data based on research carried out by contractor: Applied Research Associates)

Y = # Core Columns Damaged

WTC Sensitivity Analysis

WTC: 1-FAT Versus Orthogonal Designs
$(k = 13, n = 17)$

$$(k = 13, n = 17^-)$$

2^{13-9} wcp

Orthogonal Fractional Factorial Design $(k=13, n=17)$

Run	X1	X2	X3	X4	X5	X6	X7	X8	X9	X10	X11	X12	X13
1	1	1	-1	1	1	-1	-1	1	1	-1	-1	-1	1
2	1	1	-1	-1	-1	1	1	-1	-1	1	1	-1	1
3	1	-1	-1	1	-1	1	-1	1	-1	1	-1	1	-1
4	1	-1	-1	-1	1	-1	1	-1	1	-1	1	1	-1
5	-1	1	1	1	1	1	-1	-1	-1	-1	1	-1	-1
6	-1	1	1	-1	-1	1	1	1	1	-1	-1	-1	-1
7	-1	-1	1	1	-1	-1	-1	-1	1	1	1	1	1
8	-1	-1	1	-1	1	1	1	1	-1	-1	-1	1	1
9	0	0	0	0	0	0	0	0	0	0	0	0	0
10	1	1	1	1	-1	1	1	-1	1	-1	-1	1	-1
11	1	1	1	-1	1	-1	-1	1	-1	1	1	1	-1
12	1	-1	1	1	1	1	-1	1	-1	1	-1	-1	1
13	1	-1	1	-1	-1	1	1	1	1	-1	1	-1	1
14	-1	1	-1	1	-1	-1	1	-1	-1	-1	1	1	1
15	-1	1	-1	-1	1	1	-1	-1	1	1	-1	1	1
16	-1	-1	-1	1	1	1	1	1	1	1	1	-1	-1
17	-1	-1	-1	-1	-1	-1	-1	-1	-1	-1	-1	-1	-1

"Figure" 1.4 Data from 2^{13-9} (with center point) orthogonal experiment design
for engine/core-column impact study

WTC Sensitivity Analysis

Main Effects Plot

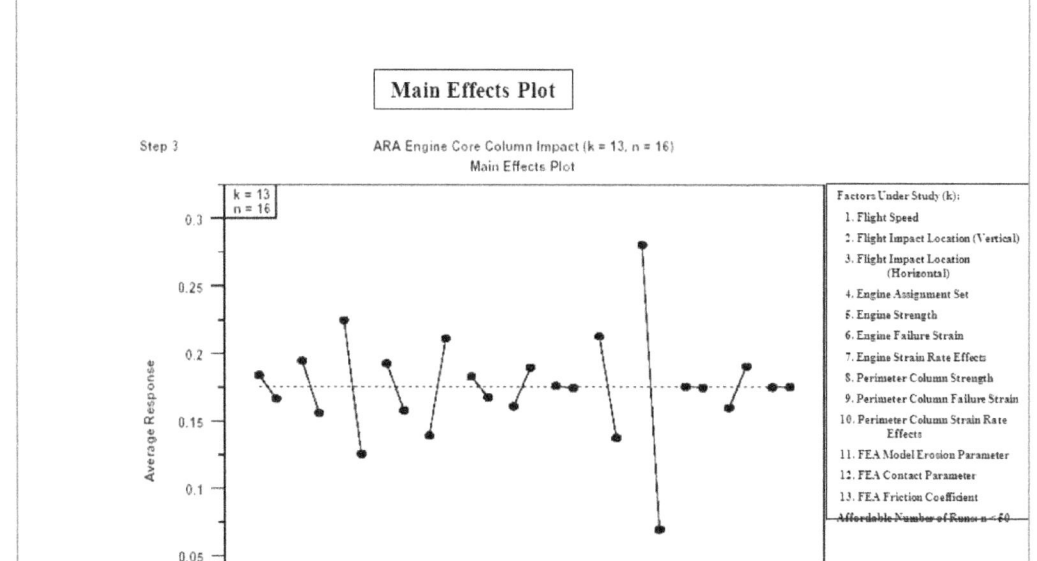

Step 3

ARA Engine Core Column Impact (k = 13, n = 16)
Main Effects Plot

Factors Under Study (k):

1. Flight Speed
2. Flight Impact Location (Vertical)
3. Flight Impact Location (Horizontal)
4. Engine Assignment Set
5. Engine Strength
6. Engine Failure Strain
7. Engine Strain Rate Effects
8. Perimeter Column Strength
9. Perimeter Column Failure Strain
10. Perimeter Column Strain Rate Effects
11. FEA Model Erosion Parameter
12. FEA Contact Parameter
13. FEA Friction Coefficient

Affordable Number of Runs: n < 50

1. Problem Classification

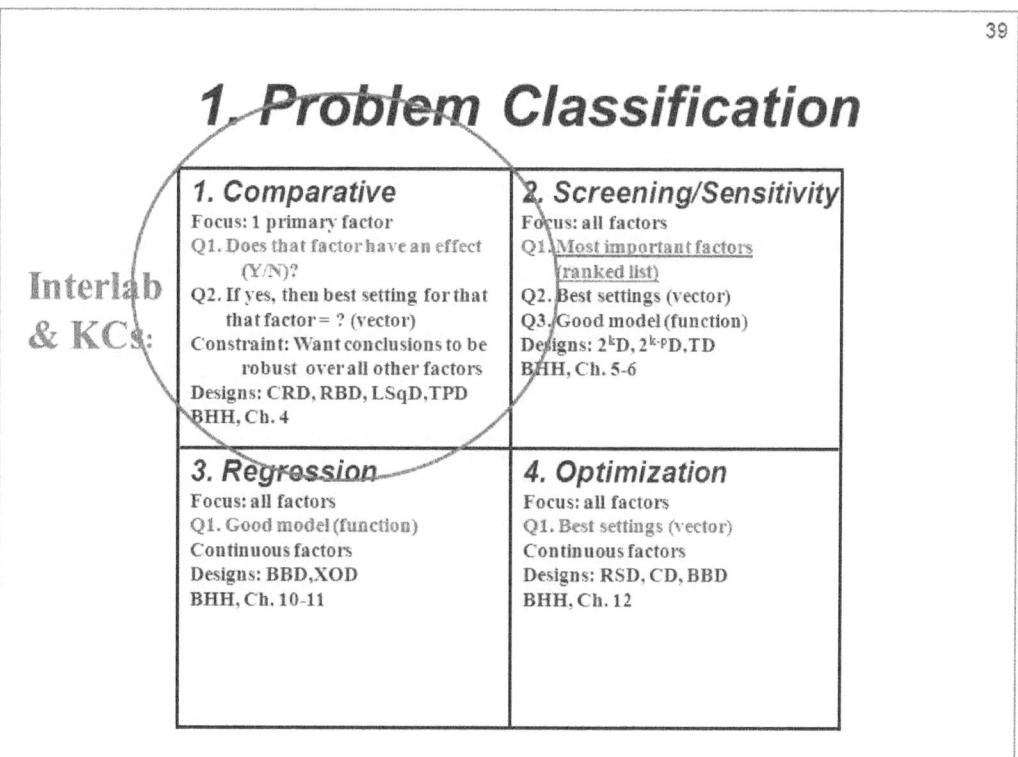

1. Comparative	2. Screening/Sensitivity
Focus: 1 primary factor Q1. Does that factor have an effect (Y/N)? Q2. If yes, then best setting for that that factor = ? (vector) Constraint: Want conclusions to be robust over all other factors Designs: CRD, RBD, LSqD, TPD BHH, Ch. 4	Focus: all factors Q1. Most important factors (ranked list) Q2. Best settings (vector) Q3. Good model (function) Designs: 2^kD, $2^{k-p}D$, TD BHH, Ch. 5-6
3. Regression	4. Optimization
Focus: all factors Q1. Good model (function) Continuous factors Designs: BBD, XOD BHH, Ch. 10-11	Focus: all factors Q1. Best settings (vector) Continuous factors Designs: RSD, CD, BBD BHH, Ch. 12

Interlab & KCs:

Strain 1

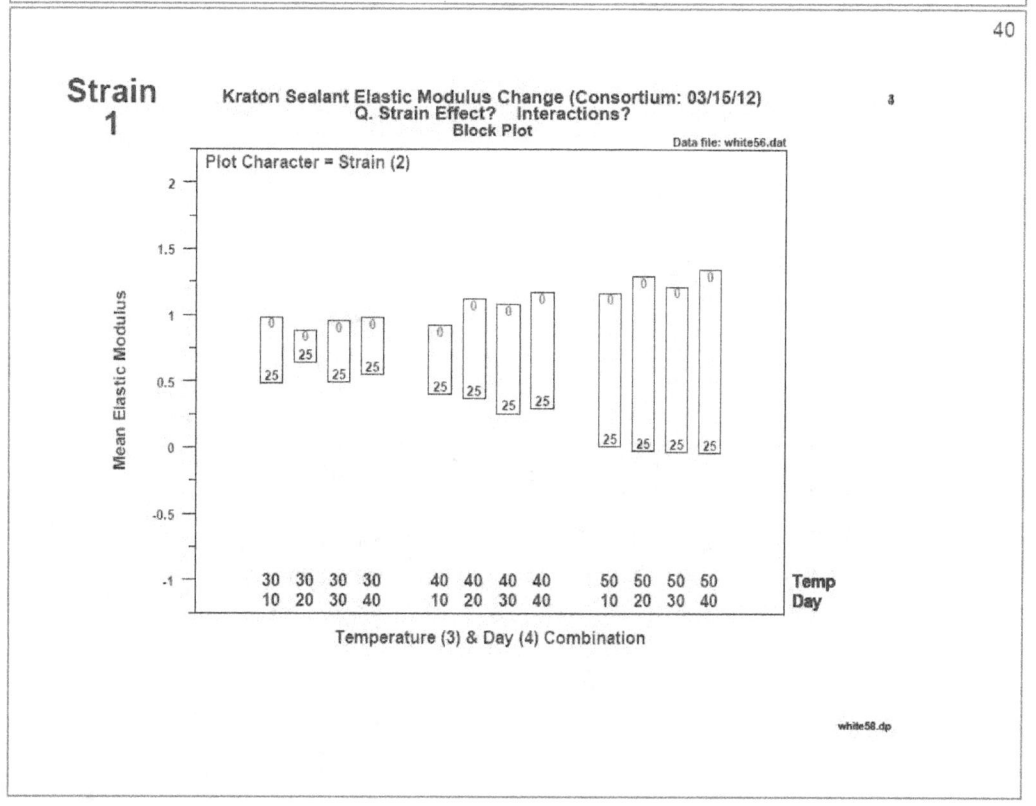

Kraton Sealant Elastic Modulus Change (Consortium: 03/15/12)
Q. Strain Effect? Interactions?
Block Plot

Temp 1

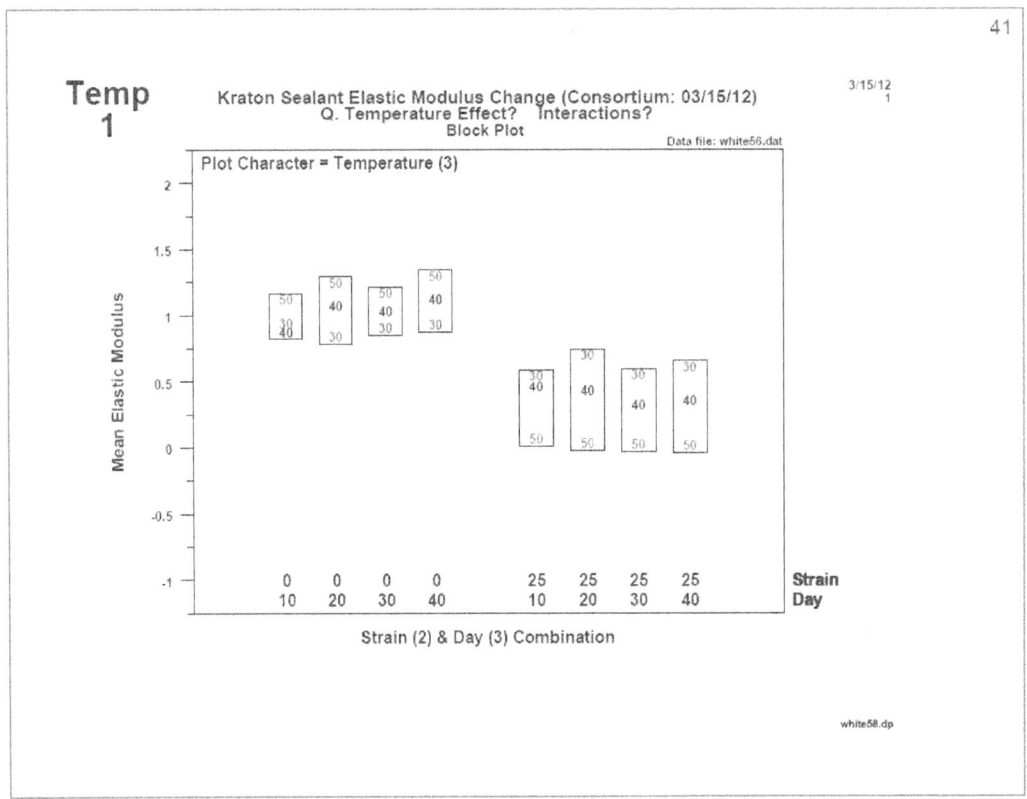

Kraton Sealant Elastic Modulus Change (Consortium: 03/15/12)
Q. Temperature Effect?   Interactions?
Block Plot

3/15/12
1

Data file: white56.dat

Plot Character = Temperature (3)

white58.dp

---

1. Problem Classification

Interlab & KCs:

1. Comparative	2. Screening/Sensitivity
If I change from lab A to lab B, does that change the TC?	What are the most important factors (and interactions) which affect TC?
3. Regression	**4. Optimization**
What is a good model so that TC may be predicted via simulation?	What are the system parameter settings that optimizes the accuracy of TC?

1. Problem Classification

Usual DEX Sequence pre-Comparative:
    1. Define Scope
    2. 1 Lab, many reps
    3. 1 Lab, Sensitivity Analysis Design
    4. Multi-Lab, Comparative Design

4. DEX Problem Translation

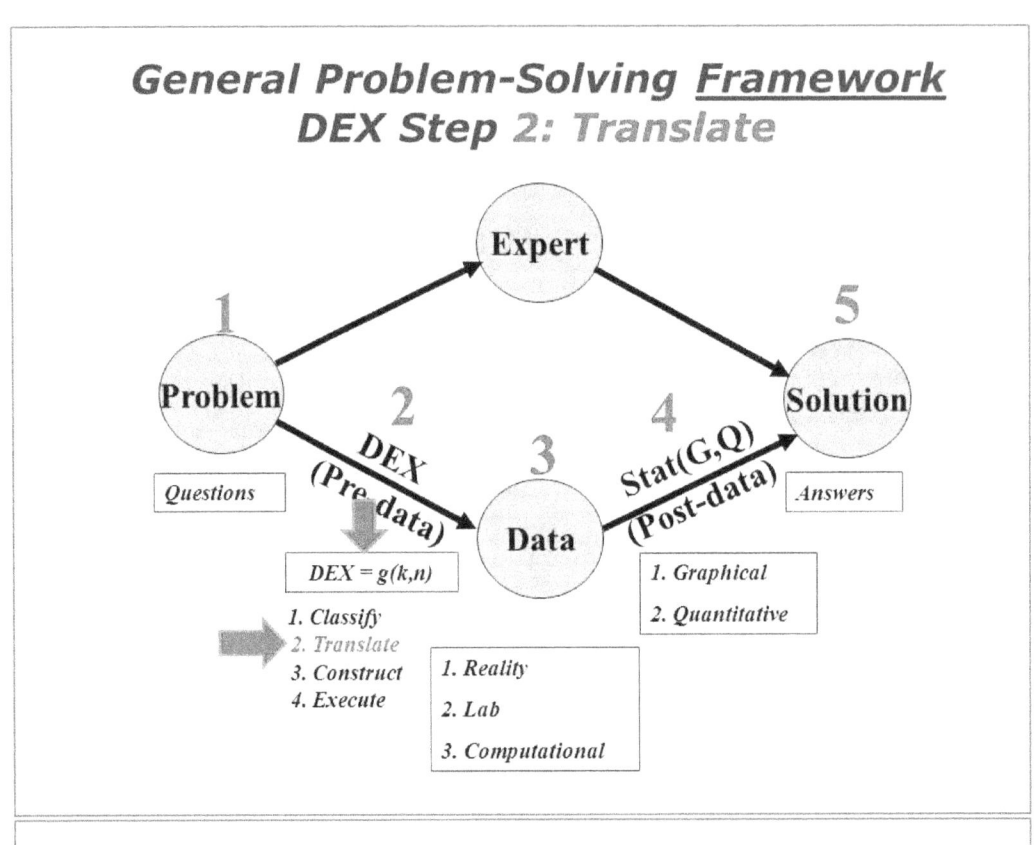

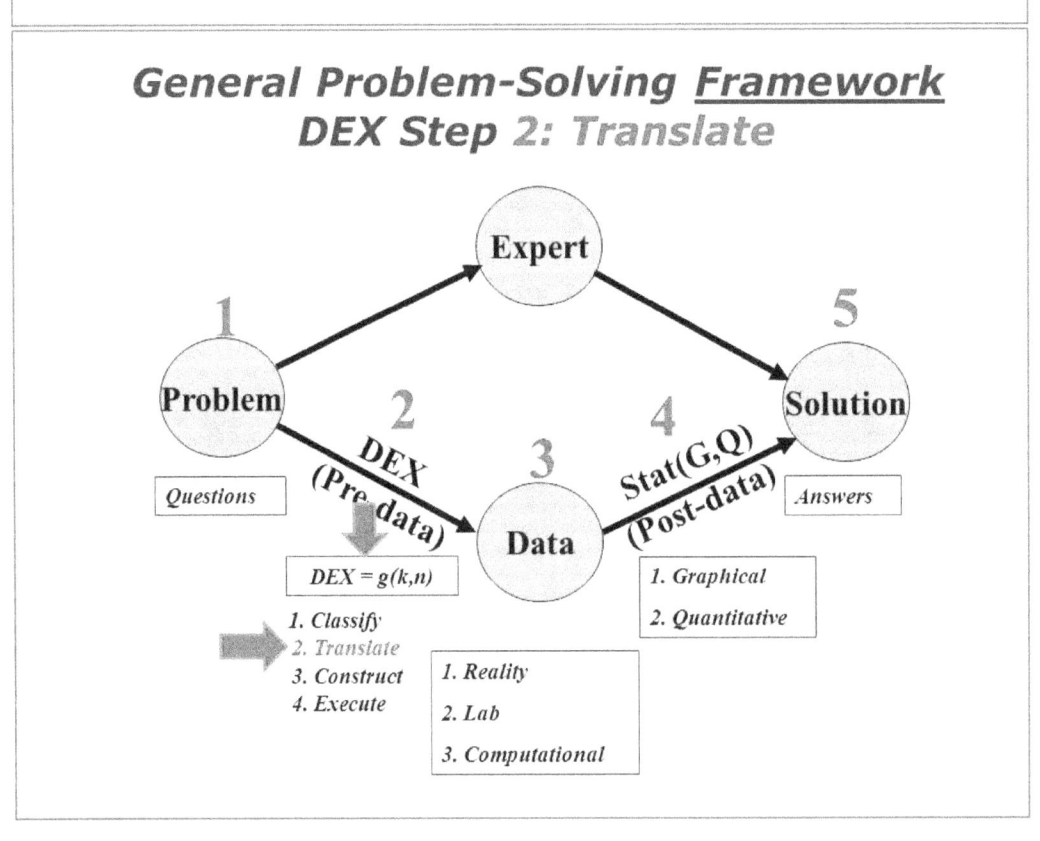

2. Problem Translation:
Minimal Info to Start the DEX Process

Specificity is key ...

1.   Title:

2.   Problem/Question:

3.   Response:

4.   Number k of Factors to Vary:

5.   Number n of Runs Affordable:

2. Problem Translation:
Minimal Info to Start the DEX Process

> For your project/problem:
>
> 1. Title = _____
>
> _____
> _____
>
> 2. Problem/Question = _____
>
> _____
> _____
>
> 3. Response Y   = _____
> 3. Number k of Factors to Vary   = _____
> 4. Sample Size n   = _____

DEX Worksheet

Experiment Design Worksheet					Date:
1. Project/Problem Title:					
2. Researcher:					
3. Project Background & Importance:					
4. General Project Question:					
5. Specific Project Question (This Experiment Only):					
6. (Generic) Stat Goal(s):					
7. Scope of Conclusions:					

DEX Essentials — Generic Stat Model: $Y = f(X1, X2, ..., Xk) + e$

8. Response Variable Y:
9. Current Typical Value for Y:
10. Project Target Value for Y:
11. Project Min. Eng. Signif. Diff. for Y:
12. Project Min. Eng. Residual SD for Y:
13. Run Time & Cost per Observation:
14. Total Available Experiment Time & Budget:
15. Constraint: Max Affordable Number of Runs:
16. Number of Factors to Vary/Investigate:
17. Factors & Factor Levels

Factor	Cont or Disc	Range	#Levels DEX1	#Levels DEX2	Mapping of Levels
X1:					
X2:					
X3:					
X4:					
X5:					
X6:					
X7:					
X8:					
X9:					
X10:					

18. General DEX Category: Comp, Scr/Sens, Regr, Optim, Ver/Fault, Unc:
19. Specific DEX:

dexworksheet.xls

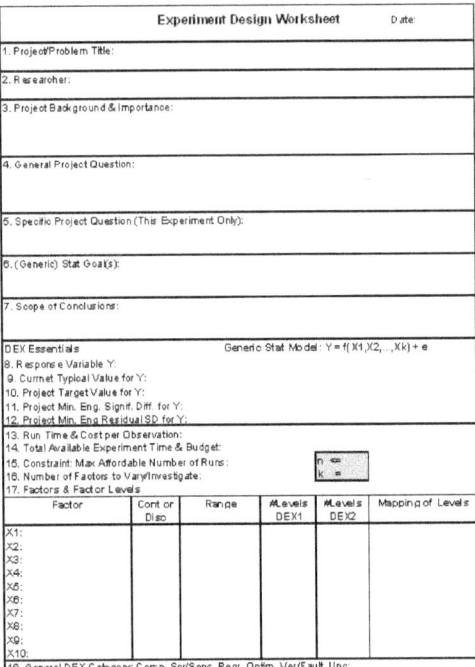

DEX Worksheet (2-page)

Date:

Experiment Design Worksheet

1. Project/Problem Title:

2. Researcher:

3. Specific Project Question (for this experiment only):

4. Project Background & Importance:

5. Specific Project Deliverables (for this experiment only):

6. (Generic) Stat Question(s) & Goal(s):

7. Scope of Conclusions:

DEX Essentials: Generic Stat Model: $Y = f(X_1, X_2, ..., X_k) + e$

8. Response Variable Y :
9. Current Typical Value for Y :
10. Project Target Value for Y :

11. Project Min. Eng. Significant D for Y :
12. Project Min. Eng. Residual SD for :

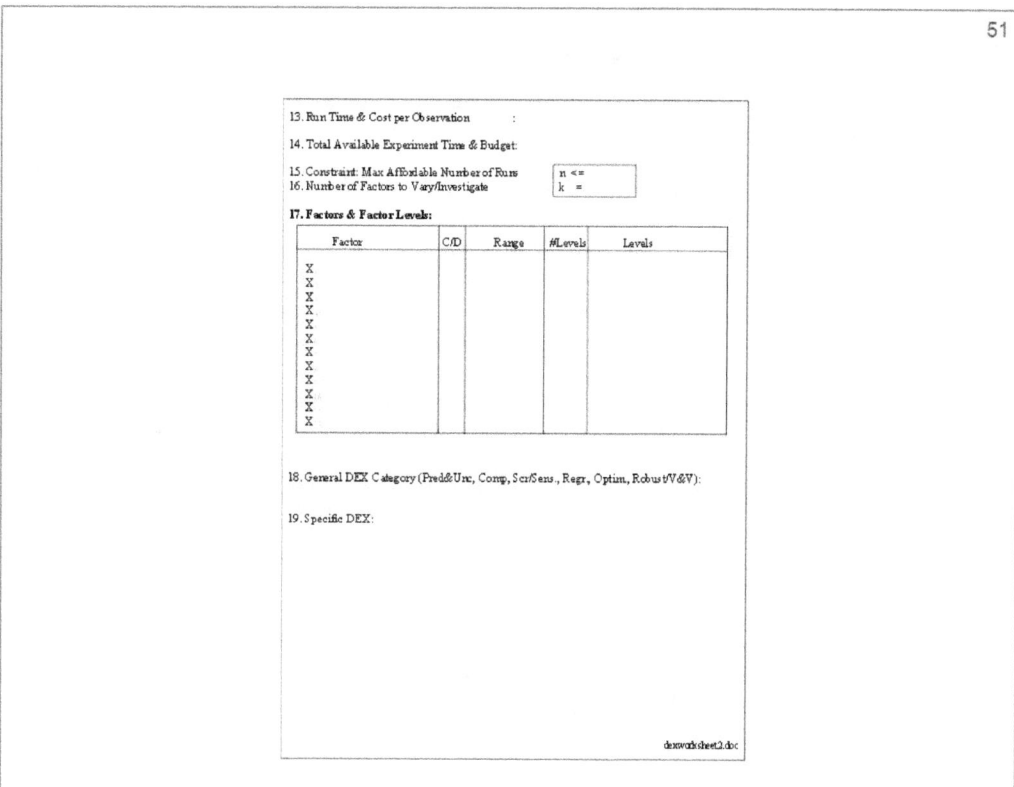

13. Run Time & Cost per Observation :

14. Total Available Experiment Time & Budget:

15. Constraint: Max Affordable Number of Runs   n <=

16. Number of Factors to Vary/Investigate   k =

17. Factors & Factor Levels:

Factor	C/D	Range	#Levels	Levels
X				
X				
X				
X				
X.				
X				
X				
X				
X				
X				
X				
X				

18. General DEX Category (Pred&Unc, Comp, Scr/Sens., Regr, Optim, Robust/V&V):

19. Specific DEX:

dexworksheet2.doc

Factors Affecting Thermal Conductivity
$Y = f(X1, X2, X3, X4, ..., Xk)$

1. Lab
2. Material
3. Q
4. L = Thickness
5. A = Area
6. Del = Temp Difference
7. Temp
8. Bulk Density
9. Rel Humidity
10. Clamp Force
11. ...
12. Device
13. Procedure
14. Pre-Conditioning
15. Ramp Up
16. Sensor Positioning
17. Operator
18. Day
19.
20.

(k=18,n= ...)

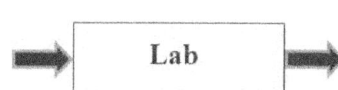

Lab

Factor Categories
Ishikawa Diagram

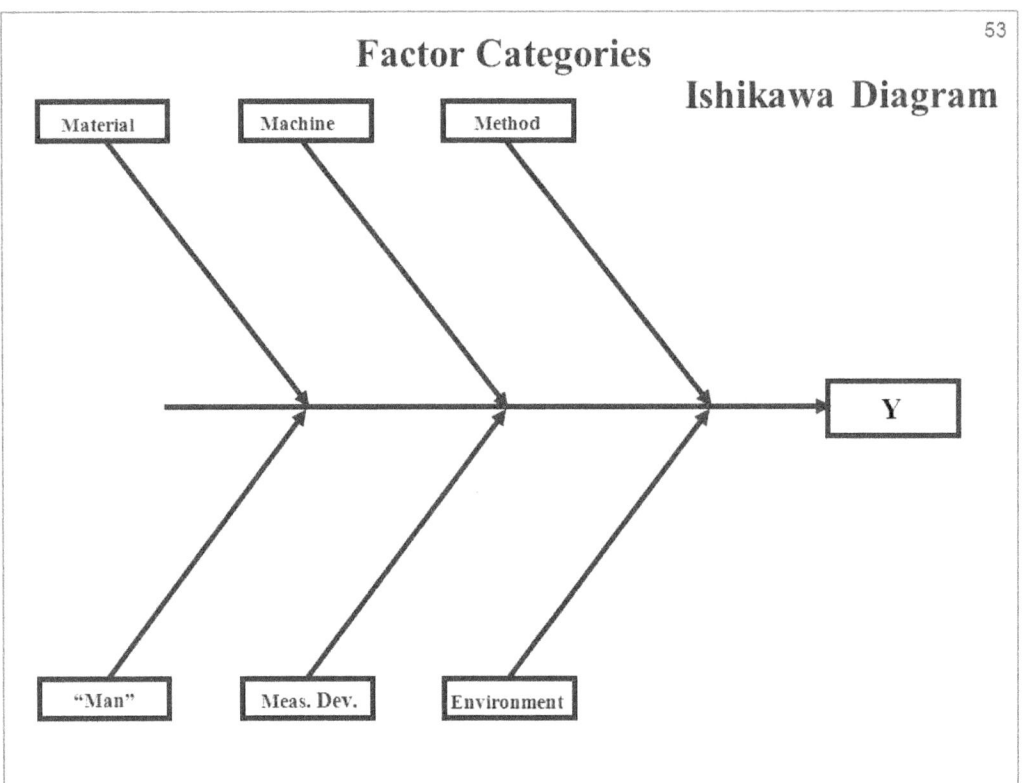

Material   Machine   Method

Y

"Man"   Meas. Dev.   Environment

---

2. Problem Translation:
Minimal Info to Start the DEX Process

1. Title:

2. Problem/Question:

3. Number k of Factors to Vary:

4. Response:

5. Number n of Runs Affordable:

2. Problem Translation:

Choices ...

1. Number of Factors k

2. Number of Levels li

3. Which Levels

(Note: DEX Principle: Sample Where the Variation Is)

4. Number of Reps r

5. DEX Principles & Techniques

DEX Principles & Techniques

Principles	Techniques
Construct Efficient Designs	Elicit Dominant Project Goal(s)
Construct Effective Designs	Elicit Project Scope & Constraints
Infer about Population	Randomize
Avoid Biased Factors	Randomize, Block, Balance Orthogonal
Maximize Test Sensitivity	Balance
Estimate All Model Parameters	n(distinct) >= k+1
Allow for Expanded Model	Record Additional (e.g., Ambient) Variables
Estimate Sigma Model-Free	Replicate
Estimate Main Effects & Int.	Full Factorial or High Res. Fractional Factorial Designs
Save $	Pilot Study, Fract. Fact. Designs
Make Conclusions Robust	Design in Many Robustness Factors
Assess Drift	Controls, Replicate across Time
Avoid Bias from Drift	Drift-Reducing Designs
Avoid Confounding	Full Factorial or High Res. Fractional Factorial Designs
Minimize SD(Estimates)	Sample Where Variation Is
Realistically Sample the Process	Design in Vicissitudes (Multiple Sets)
Reduce Effect Uncertainty	Youden Pairs for Homogeneity
Assess Repeatability	Replication
Assess Reproducibility	Multiple Sets (Across Days)

dexprintech.dp

DEX Principles & Techniques

Principles	Techniques
Construct Efficient Designs	Elicit Dominant Project Goal(s)
Construct Effective Designs	Elicit Project Scope & Constraints
Infer about Population	Randomize
Avoid Biased Factors	Randomize, Block, Balance Orthogonal
Maximize Test Sensitivity	Balance
Estimate All Model Parameters	n(distinct) >= k+1
Allow for Expanded Model	Record Additional (e.g., Ambient) Variables
Estimate Sigma Model-Free	Replicate
Estimate Main Effects & Int.	Full Factorial or High Res. Fractional Factorial Designs
Save $	Pilot Study, Fract. Fact. Designs
Make Conclusions Robust	Design in Many Robustness Factors
Assess Drift	Controls, Replicate across Time
Avoid Bias from Drift	Drift-Reducing Designs
Avoid Confounding	Full Factorial or High Res. Fractional Factorial Designs
Minimize SD(Estimates)	Sample Where Variation Is
Realistically Sample the Process	Design in Vicissitudes (Multiple Sets)
Reduce Effect Uncertainty	Youden Pairs for Homogeneity
Assess Repeatability	Replication
Assess Reproducibility	Multiple Sets (Across Days)

dexprintech.dp

Important DEX Issues

1. Equivalance
2. Effect
3. Sample & Population
4. Randomization
5. Coverage
6. Balance
7. Orthogonality
8. Scope & Robustness
9. Confounding
10. Blocking
11. Replication
12. Sample Size n

1 & 2. Equivalance & Effect

2 labs are equivalent if ...
The effect due to laboratory is ...

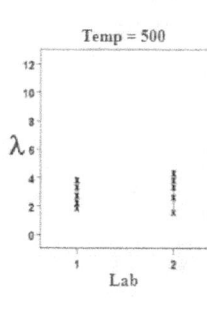
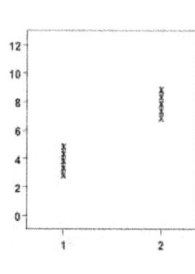
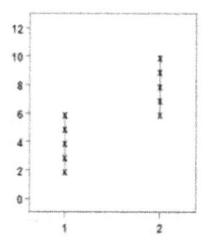
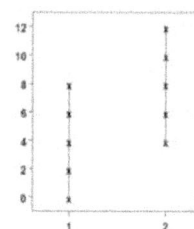
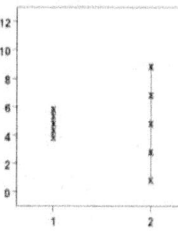
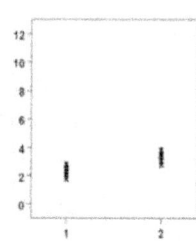

1 & 2. Equivalance & Effect

An effect is ...

1. By default, "effect" means shift in location.

2. Our ability to detect an effect depends on the intrinsic (within-lab) variability of the data σ_i

3. "Effect" could also mean shift in variation (within lab $\sigma_{i)}$

4. We <u>conclude</u>: "a factor has an effect" by computing a minimum <u>statistical</u> significant difference (via statistical hypothesis testing). Of equal importance, is the minimum engineering significant difference.

1 & 2. Equivalance & Effect

Equivalence ...

1. By default, 2 labs are "equivalent" if they are statistically identical in location.

2. Our ability to detect equivalence depends on the intrinsic (within-lab) variability of the data σ_i

3. Equivalence could also mean statistically identical in variation (within lab σ_i)

4. We <u>conclude</u>: "2 labs are equivalent" by computing and comparing location estimates (via statistical hypothesis testing, e.g., t, ANOVA). Of additional interest is whether the location differences exceed the "minimum engineering significant difference" (scientist-provided).

266

1 & 2. Equivalance & Effect

Hot Plate:
   Min Eng Significant Difference = ...

Pipe:
   Min Eng Significant Difference = ...

---

3. Sample & Population

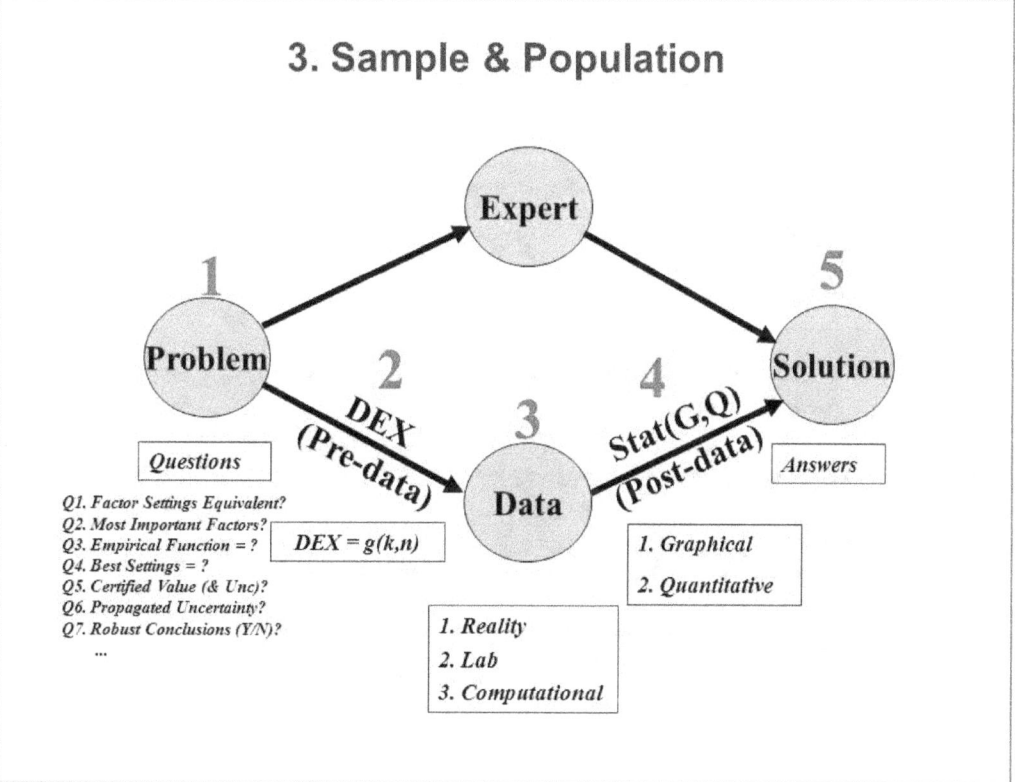

3. Sample & Population

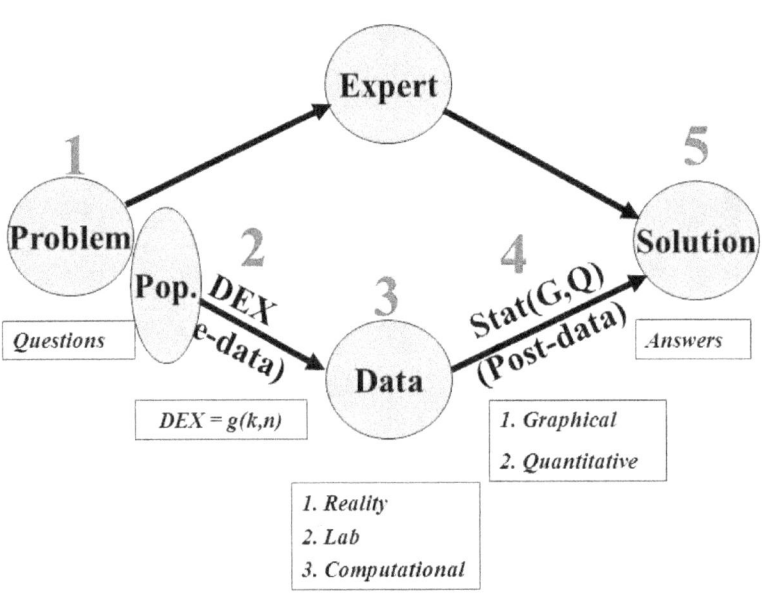

Expert

1
Problem

Pop. DEX
e-data)

Questions

2

DEX = g(k,n)

3
Data

4
Stat(G,Q)
(Post-data)

5
Solution

Answers

1. *Graphical*
2. *Quantitative*

1. *Reality*
2. *Lab*
3. *Computational*

3. Sample & Population

$$Y = f(X1, X2, X3, \ldots, Xk)$$

?

What is the population {...} of my thermal conductivity measurements?
What factors define (my) population? (Do I have a population?)
What scope (robustness) do I want my conclusions
    to be valid over?
 What is the population of materials, operators, devices,
    procedures, envirnoments, etc.

Population {...}
N =

Sample {...}
n =

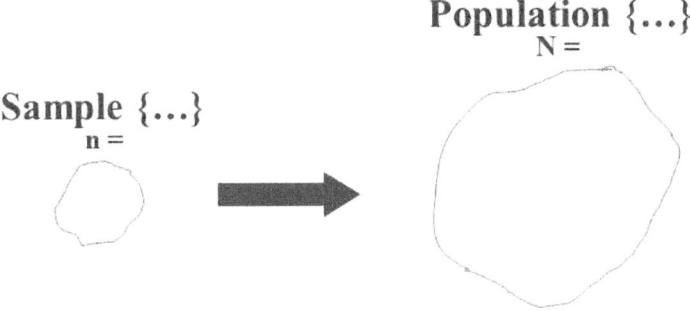

3. Sample & Population

$$Y = f(X1, X2, X3, \ldots, Xk)$$

?

What is the population {...} of my thermal conductivity measurements?
What factors define (my) population? (Do I have a population?)
What scope (robustness) do I want my conclusions
   to be valid over?
 What is the population of materials, operators, devices,
   procedures, envirnoments, etc.

Population {...}
N =

Sample {...}
n =

Representative →

67

---

3. Sample & Population

$$Y = f(X1, X2, X3, \ldots, Xk)$$

What is ... asurements?
What f... on?)
What s...
   to b...
What i...
   pro...

If the sample is not statistically representative of the population (or if there is no population), then we are doing
   1. summarization, rather than
   2. inference

Sample {...}
n =

Representative →

$\bar{y}$, s, histogram
Statistics

μ, σ, pdf
Parameters

68

269

4. Randomization

Randomization: A sampling technique--
   1. Non-predictability
   2. Every element in the population has an equal chance of being drawn
   3. Basis for statistical inference and estimation
   4. For a material factor, helps assure lab factor bias protection
   5. For a time factor, helps provide drift protection
   6. Some aspect of randomization should be in every experiment

Population {…}
N =

Sample {…}
n =

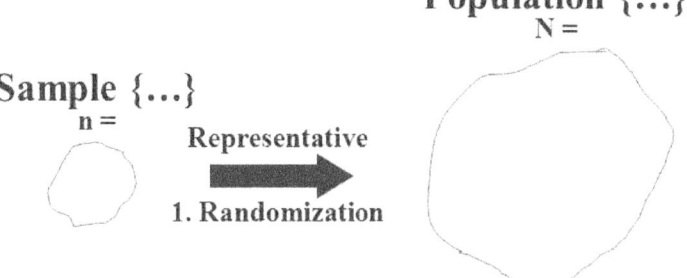

Representative

1. Randomization

69

---

4. Randomization

Randomization: A sampling technique--
   1. Non-predictability
   2. Every element in the population has an equal chance of being drawn
   3. Basis f
   4. For a
   5. For a ti
   6. Some a

Advantages of balance:
   1. Helps assure representativeness
   2. Easy to carry out—even for complicated multi-factor populations

Sample {…}
n =

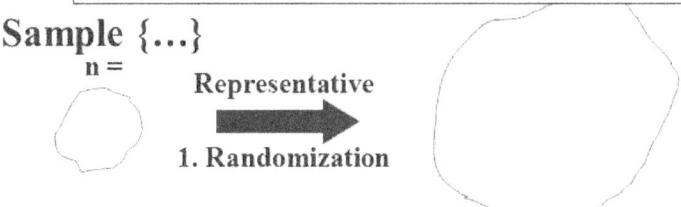

Representative

1. Randomization

70

5. Coverage

A factor has coverage if every level of that factor has at least one observation collected at it.

Population {…}
N =

Sample {…}
n =

Representative

1. Randomization
2. Balance

71

---

5. Coverage

A factor has coverage if every level of that factor has at least one observation collected at it.

Advantages of coverage:
1. Helps assure representativeness
2. Helps embroaden scope
3. Data-based conclusions must have data

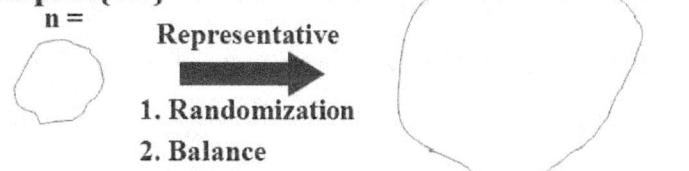

n =

Representative

1. Randomization
2. Balance

72

6. Balance

A factor is balanced if every level of that factor occurs the same
    number of times.
A design is balanced if every factor is balanced.

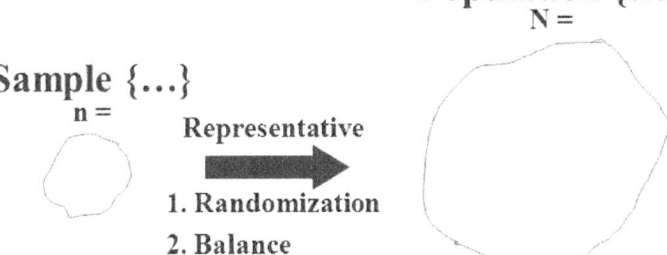

Population {...}
N =

Sample {...}
n =

Representative

1. Randomization
2. Balance

73

6. Balance

A factor is balanced if every level of that factor occurs the same
    number of times.

A de

Sam

Advantages of balance:
    1. Helps assure representativeness
    2. If a factor effect actually exists,
       then balance maximizes our
       ability to statistically conclude:
       "this factor has an effect"
       (optimizes a t-test)

2. Balance

74

6. Balance

t-test

To determine if a statistically significant difference in location exists, the t statistic is as follows:

$$t_{stat} = \frac{\bar{y}_2 - \bar{y}_1}{s\sqrt{\frac{1}{n_1} + \frac{1}{n_2}}}$$

t_{stat} is compared to t_{crit}

where t_{crit} is theoretical value taken from $t_{n1+n2-2}$ distribution

6. Balance

For example, if total n = 6
      (2 labs and 3 TC measurements at a fixed temp per lab)
      $\bar{y}_1 = 6$
      $\bar{y}_2 = 8$
Is there a statistically significant shift in location?

n1	n2	t_{stat}	$t_{crit}=t_{4, 0.975}$	Shift?
3	3	$t_{stat}=\frac{\bar{y}_2-\bar{y}_1}{s\sqrt{\frac{1}{n_1}+\frac{1}{n_2}}}=\frac{8-6}{0.85\sqrt{\frac{1}{3}+\frac{1}{3}}}=$ 2.88	2.78	Yes
4	2	$t_{stat}=\frac{\bar{y}_2-\bar{y}_1}{s\sqrt{\frac{1}{n_1}+\frac{1}{n_2}}}=\frac{8-6}{0.85\sqrt{\frac{1}{4}+\frac{1}{2}}}=$ 2.72	2.78	No
5	1	$t_{stat}=\frac{\bar{y}_2-\bar{y}_1}{s\sqrt{\frac{1}{n_1}+\frac{1}{n_2}}}=\frac{8-6}{0.85\sqrt{\frac{1}{5}+\frac{1}{1}}}=$ 2.15	2.78	No

The t-test is sensitivity to imbalance

7. Orthogonality

2 factors are orthogonal if each factor itself is balanced and if every pairwise combination of the levels of the 2 factors is also balanced (occurs the same number of times)

A design is balanced if every factor is balanced, and every pair of factors is balanced.

Population {…}
N =

Sample {…}
n =

Representative

1. Randomization
2. Balance
   Orthogonality

77

---

7. Orthogonality

2 factors are orthogonal if each factor itself is balanced and if every combination of the pairs of levels of the 2 factor

is a

A de

pa

Sam

Advantages of orthogonality:

1. *Helps assure representativeness*
2. *Allows effect estimates to be independent of one another.*
3. *Allows interaction effects to be detected and estimated (if numerically possible)*
4. *Orthogonal fractional factorial designs are extremely important and cost-effective for carrying out Sensitivity Analyses.*

78

7. Orthogonality

(k=5, n=16) $2^{(5-1)}$

X1	X2	X3	X4	X5
-	-	-	-	+
+	-	-	-	-
-	+	-	-	-
+	+	-	-	+
-	-	+	-	-
+	-	+	-	+
-	+	+	-	+
+	+	+	-	-
-	-	-	+	-
+	-	-	+	+
-	+	-	+	+
+	+	-	+	-
-	-	+	+	+
+	-	+	+	-
-	+	+	+	-
+	+	+	+	+

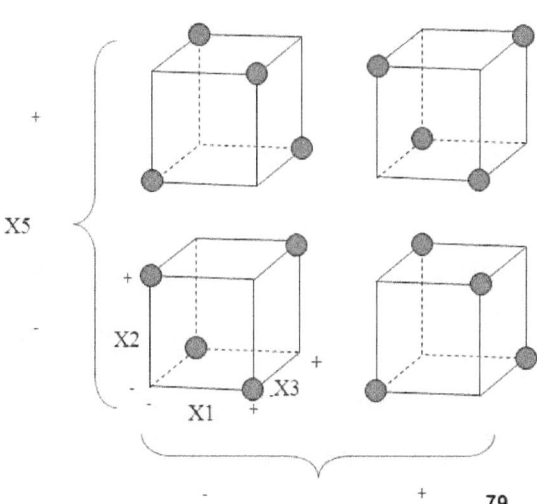

79

---

7. Orthogonality

Orthogonal Design Balance
(Balanced in both 1 and 2 dimensions)
(k=5,n=16)

For 2^{5-1} Design:

All k:

$$\frac{8 \quad\quad 8}{{}_{-}\quad X_i \quad {}^{+}}$$

All $\binom{k}{2}$: X_j

```
      +  ┌─────────┐
         │ 4     4 │
         │         │
         │ 4     4 │
      -  └─────────┘
         -    X_i    +
```

275

7. Orthogonality

(k=5, n=8) $2^{(5-2)}$

X1	X2	X3	X4	X5
-	-	-	+	+
+	-	-	-	+
-	+	-	-	-
+	+	-	+	-
-	-	+	+	-
+	-	+	-	-
-	+	+	-	+
+	+	+	+	+

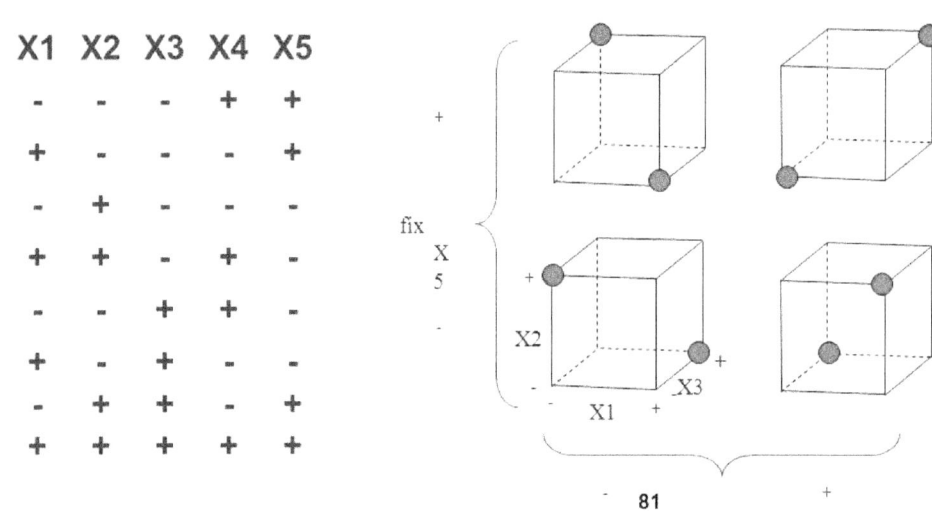

81

7. Orthogonality

(k=5, n=6) 1-FAT (Non-orthogonal)

X1	X2	X3	X4	X5
-	-	-	-	-
+	-	-	-	-
-	+	-	-	-
-	-	+	-	-
-	-	-	+	-
-	-	-	-	+

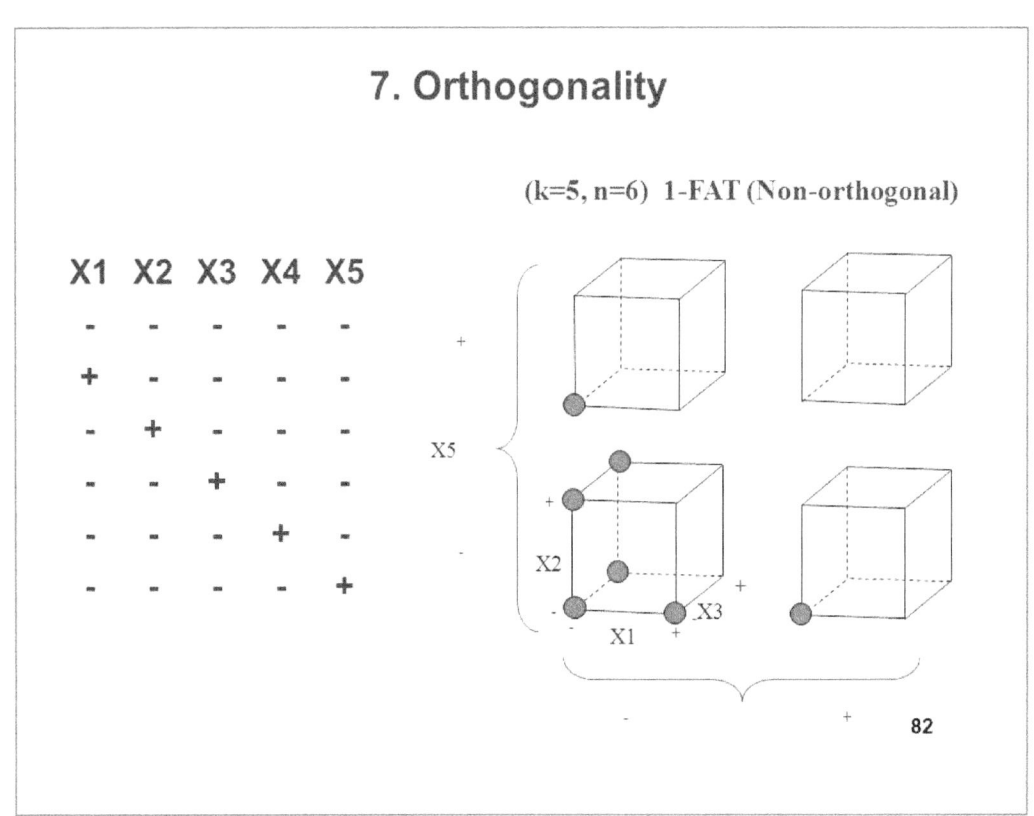

82

276

8. Scope & Robustness

$$Y = f(X1, \quad X2, \quad X3, \dots, Xk)$$

$$Y = f(Lab, \; Temp, \; Mat, \dots, Xk)$$

Primary   Robustness
factor   factors

Q. What robustness factors are important
components in your <u>population</u>-of-interest?

Population {...}
N =

Sample {...}
n =

Representative

---

8. Scope & Robustness

$$Y = f(X1, \quad X2, \quad X3, \dots, Xk)$$

$$Y = f(Lab, \; Temp, \; Mat, \dots, Xk)$$

> *To expand the scope of the experiment,
> data can/must be collected / designed
> (passive / active) over a range of
> additional factors.*
>
> *From a DEX point of view, these
> additional (= robustness) factors
> must also be handled with care
> (balance, coverage, etc.)*

Q. V

Sar

8. Scope & Robustness

What are Good Robustness Factors?

$$Y = f(X1, X2, X3, \ldots, Xk)$$

?

5M's & E

1. Material
2. Machine/Device
3. Method/Protocol
4. "Man"/Operator
5. Measuring Device
6. Environment

---

8. Scope & Robustness

Material	Machine	Method

Ishikawa Diagram

Y

"Man"	Meas. Dev.	Environment

8. Scope & Robustness

Robustness Factors are the Key ...

1. The robustness factors define our <u>scope</u> of conclusions

    about the primary factor.

        Choice/decision for you: What scope do you want--narrow or broad?

        If fix all robustness factors, then scope = narrow

        Your data-based conclusions are limited to where you collect data

2. Ideally, the robustness factors do <u>not</u> interact with

    the primary factor.

        If not interact, then have robustness.

        If do interact, then have non-robustness (& insight)

---

8. Scope & Robustness

Robustness Factors: What we do (DEX) ...

1. Let R.F. float: do not vary them (and do not record them)
2. Let R.F. float: do not vary them (but do record them)
3. Fix all Robustness Factors
4. Vary/control one Robustness Factor
5. Vary/control two Robustness Factors
6. Vary/control "all" Robustness Factors

8. Scope & Robustness

Robustness Factors: What we do (DEX) &
Consequences on Scope ...

1. Let them float: do not vary them (and do not record them) unknown/terrible

2. Let them float: do not vary them (but do record them) poor

3. Fix them all ok but narrow

4. Vary/control one of them better

5. Vary/control two of them better still

6. Vary/control "all" of them best/broad

---

8. Scope & Robustness

Us vs. Nature: Choosing the
Correct Robustness Factors ...

Our model: $Y = f_{SE}(X_1, X_2, X_3, ..., X_k)$

Nature's model: $Y = f_N (X_1, X_2, X_3, ..., X_k, X_{k+1} ..., X_N)$

8. Scope & Robustness

Thermal Conductivity Models

Our Model: Y　　　= f(X1, X2)(k=2,n=20)

Y = Thermal Conductivity
X1 = Laboratory　　(5 levels: 1, 2, 3, 4, 5)
X2 = Temperature　(4 levels: 100, 200, 300, 400)

Nature's Model: Y　 = f(X1, X, X3=?,n=20)

Y = Thermal Conductivity
X1 = Laboratory　　(5 levels: 1, 2, 3, 4, 5)
X2 = Temperature　(4 levels: 100, 200, 300, 400)
X3 = ?
X4 = ?
...

8. Scope & Robustness

Us vs. Nature: Choosing the Correct Robustness Factors ...

Our model:　　　$Y = f_{SE}(X_1, X_2, X_3, ..., X_k)$

Nature's model: $Y = f_N (X_1, X_2, X_3, ..., X_k, X_{k+1} ..., X_N)$

There is no substitute (or recovery) from the scientist/

engineer choosing an <u>incomplete</u> set of robustness factors.

This requires skillful S/E expertise.

9. Confounding

- *Confounding occurs in a design when levels of one factor are directly correlated with levels of another factor*

Temp = 500

Lab	Material Type	Result
1	1	7.24
1	1	8.15
1	1	6.98
2	2	10.23
2	2	11.05
3	2	10.78

Lab - Material Type Confounding: The observed difference in results cannot be directly attributed to either Lab or Material Type

9. Confounding

- *Confounding occurs in a design when levels of*

Confounding is a curse to any experiment inasmuch as valid, crisp and unambiguous results cannot be drawn.

Confounding factors may not always be obvious:
- *Material, Material Type*
- *Environmental factors*
- *Time (devices may drift over time)*
- *Learning effect/run order*
- *Local power sources drifting*

10. Blocking

- *A design technique applied to robustness factors to assure <u>anti-confounding</u>*
- *A robustness factor is a blocking factor if each & every level of the robustness factor has each & every level of the primary factor occurring the same number of times (within-block balance)*

Y = mpg
X1 = Additive (4): A, B, C, D
X2 = Car (4): I, II, III, IV
X3 = Driver (4): 1,2,3,4   (k=1+2, n=16)

Latin Square Design

		Car		
	I	II	III	IV
D r i v e r 1	A	B	C	D
2	A	B	C	D
3	A	B	C	D
4	A	B		D

		Car		
	I	II	III	IV
D r i v e r 1	B	D	A	C
2	C	C	B	D
3	A	B	D	B
4	D	A	C	A

11. Replication:
How many reps are enough?

$$n \text{ (theoretical)} >= (2\sigma/T)^2 = g(\sigma, T)$$

This n depends on data variation (σ) and
*   scientist-desired tolerance (T)*

➡ *If collect these many reps n, then you will be 95% sure that the sample mean from the resulting n observations will be within +-T of the true (unknown) population mean μ. (via Central Limit Theorem)*

12. Sample Size n:

$$Y = f(X1, X2, X3, \ldots, Xk)$$

n will grow as number of robustness factors grow

n (theoretical) >= k+1

n (theoretical) >= $(2*\sigma/T)**2 = g(\sigma, T)$

n (max) is dictated by scientist \$/time

4 Ways to Reduce n:
  1. Reduce the number of factors (but poorer scope)
  2. Reduce the number of levels   (but poorer scope)
  3. Reduce the number of reps   (but poorer $\hat{\sigma}$)
  4. Orthogonal fractional factorial design for the robustness factors (but lose higher-order interaction info)

Null Designs

6. DEX Criteria & Interlab Designs

Interlab Designs = Comparative Designs

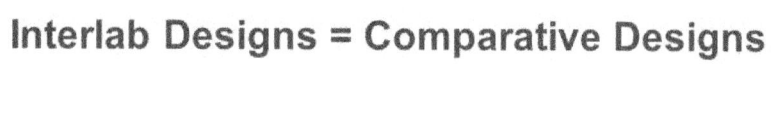

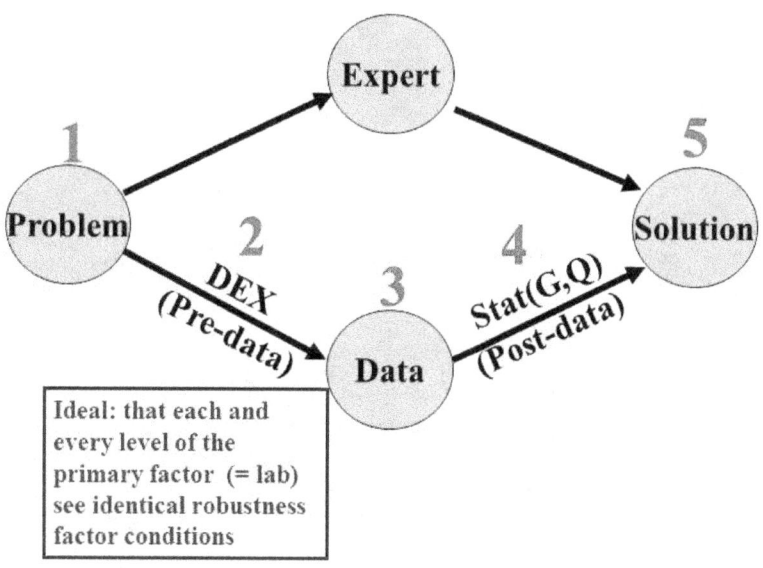

Ideal: that each and every level of the primary factor (= lab) see identical robustness factor conditions

---

Interlab Designs = Comparative Designs

$(k = \ , n = \ )$

| X6: Env. Cond ( ): |
| X5: Operator ( ): |
| X4: Protocol ( ): |
| X3: Device ( ): |
| X2: Temp ( ): |

X1: Lab	Cond : 1 2 3 4 5 6 7 8 9 10 ... n
1	
2	
3	
4	
5	

Default (General): Full Factorial Design with Replication
Q. If n = 100 (say), what design would you run?
Q. (k= ?, n = ?)
Default simple design here: (k=2, n = 2*5*2)
(Q. S.A.M Design) SRM?

2 Labs & 4 Potential Temps (k=2,n=8)

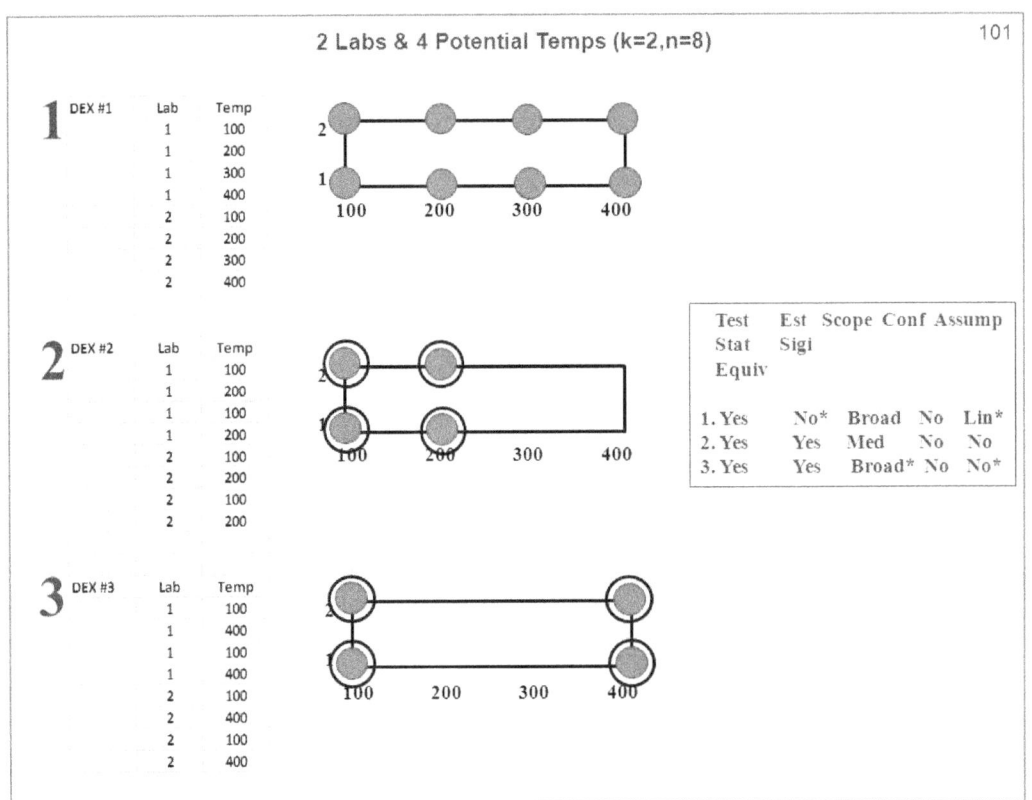

1 DEX #1

Lab	Temp
1	100
1	200
1	300
1	400
2	100
2	200
2	300
2	400

2 DEX #2

Lab	Temp
1	100
1	200
1	100
1	200
2	100
2	200
2	100
2	200

3 DEX #3

Lab	Temp
1	100
1	400
1	100
1	400
2	100
2	400
2	100
2	400

Test Stat Equiv	Est Sigi	Scope	Conf	Assump
1. Yes	No*	Broad	No	Lin*
2. Yes	Yes	Med	No	No
3. Yes	Yes	Broad*	No	No*

www.ingramcontent.com/pod-product-compliance
Lightning Source LLC
Chambersburg PA
CBHW081434170526
45166CB00008B/2200